ROUTLEDGE LIBRARY EDITIONS:
POLLUTION, CLIMATE AND CHANGE

Volume 14

CLIMATE CHANGE AND WORLD AGRICULTURE

CLIMATE CHANGE AND WORLD AGRICULTURE

MARTIN L. PARRY

LONDON AND NEW YORK

First published in 1990 by Earthscan Publications Ltd.

This edition first published in 2020
by Routledge
2 Park Square, Milton Park, Abingdon, Oxon OX14 4RN

and by Routledge
52 Vanderbilt Avenue, New York, NY 10017

Routledge is an imprint of the Taylor & Francis Group, an informa business

British Library Cataloguing in Publication Data
A catalogue record for this book is available from the British Library

ISBN: 978-0-367-34494-8 (Set)
ISBN: 978-0-429-34741-2 (Set) (ebk)
ISBN: 978-0-367-36286-7 (Volume 14) (hbk)
ISBN: 978-0-367-36291-1 (Volume 14) (pbk)
ISBN: 978-0-429-34510-4 (Volume 14) (ebk)

Publisher's Note
The publisher has gone to great lengths to ensure the quality of this reprint but points out that some imperfections in the original copies may be apparent.

Disclaimer
The publisher has made every effort to trace copyright holders and would welcome correspondence from those they have been unable to trace.

CLIMATE CHANGE AND WORLD AGRICULTURE

Martin Parry

Earthscan Publications Limited London

in association with

The International Institute for Applied Systems Analysis

•

United Nations Environment Programme

First published 1990 by
Earthscan Publications Ltd
3 Endsleigh Street, London WC1H 0DD

British Library Cataloguing in Publication Data
Parry, M.L. (Martin Lewis), *1945-*
Climate change and world agriculture.
1. Agricultural industries. Effects of climatic changes
I. Title
338.1'5
ISBN 1-85383-065-8

Production by David Williams Associates (081-521 4130)
Typeset by Bookman Ltd, Bristol
Printed and bound by Longdunn Press, Bristol

Earthscan Publications Ltd is an editorially independent and wholly owned subsidiary of the International Institute for Environment and Development (IIED).

To Charlotte Parry

CONTENTS

	Figures and Tables	viii
	Abbreviations and Acronyms	xiii
	Preface	xv
1.	The sensitivity of agriculture to climate	1
2.	Possible changes of climate	9
3.	Methods of assessing impacts of climatic change	24
4.	Effects on plants, soil, pests and diseases	37
5.	Effects on agricultural potential	61
6.	Effects on production and land use	88
7.	Implications for global food security	105
8.	Adapting to climatic change	119
9.	Conclusions	127
	Further reading	133
	Notes and References	135
	Index	151

FIGURES AND TABLES

FIGURES

Figure 1.1 Regions identified as critical zones in respect of ability to support current population

Figure 2.1 Radiative forcing, also expressed as equivalent CO_2 concentrations, resulting from four emissions scenarios selected by the IPCC

Figure 2.2 Predictions of global mean temperature change under the four IPCC scenarios

Figure 2.3 Model average of the $2 \times CO_2$ minus $1 \times CO_2$ change in surface air temperature

Figure 2.4 Changes in soil moisture due to doubling CO_2 as simulated by three high resolution models (a) and (d) Canadian Climate Centre model; (b) and (e) Geophysical Fluid Dynamics Laboratory; (c) and (f) United Kingdom Meteorological Office

Figure 2.5 Global mean combined land-air and sea-surface temperatures, 1861–1989, relative to the average for 1951–80

Figure 3.1 Schema of a) impact and b) interaction approaches in climate impact assessment

Figure 3.2 Schema of interaction approach to climate impact assessment, with ordered interactions

Figure 3.3 A hierarchy of models for integrated climate impact assessments

Figure 3.4 Schema of the IIASA/UNEP project's approach: an interactive approach to climate impact assessment with ordered interactions, interactions at each level, and some social and physical feedbacks

Figure 3.5 a) Direct and b) adjoint methods of climate impact assessment

Figure 3.6 Schema of the construction, operation and validation of a) an empirical-statistical and b) a simulation model

Figure 4.1 Typical photosynthesis response of plants to CO_2

Figure 4.2 Temperature, development and canopy expansion

Figure 4.3 Effects of temperature deviations from 1951–80 normal on: (a) changes in growing season length for crop districts 1a, 9a and a provincial average for Saskatchewan, (b) changes in maturation time for spring wheat, and (c) changes in spring wheat phase development lengths

Figure 4.4 Modelled responses of total dry matter production and grain yield of winter wheat

Figure 4.5 Temperature zones in which farm animals perform effectively

Figure 4.6 Probability of crop "failure", net loss or critical shortfall, with linear (and normally distributed) gradient of aridity or warmth

Figure 4.7 Potential number of generations of the European Corn Borer (*ostrinia nubilalis*) for a) present-day climate (1951–80) and b) a 1°C increase in mean annual temperature

Figure 4.8 Calculated boreal zone for the GISS $2 \times CO_2$ climate scenario relative to the calculated present-day zone

Figure 4.9 Potential natural vegetation map for Europe based a) on current average temperatures and precipitation and b) on average temperatures of +5°C and average precipitation of +10%

Figure 5.1 Estimations of the impacts of climatic change on the geographical extent of the US Corn Belt

Figure 5.2 Simulated North American wheat regions using the a) GISS GCM control and b) doubled CO_2 runs

Figure 5.3 Grain-maize limits in Europe under a) current climate (1951–80), and b) GISS, c) GFDL, and d) OSU equilibrium $2 \times CO_2$ climates

Figure 5.4 Safely cultivable area for irrigated rice in northern Japan under a) current climate (1951–80) and b) the GISS equilibrium 2 × CO_2 climate

Figure 5.5 Grain-maize limit under the GISS transient response Scenario A in the 1990s, 2020s, and 2050s (relative to the limit for the current climate)

Figure 5.6 Hypothetical limits for successful ripening of two crops based on temperature: (a) Grain-maize, (b) Silage maize

Figure 5.7 Present-day analogues of the GISS 2 × CO_2 climate estimated for selected regions in the IIASA/UNEP study: Saskatchewan, Iceland, Finland, Leningrad and Cherdyn regions (USSR) and Hokkaido and Tohoku districts (Japan)

Figure 5.8 Shift of 1 in 3.3 failure frequency for oats in the British Isles for 1°C increase in mean annual temperature (normals 1856–95)

Figure 5.9 Percentage change in net primary productivity relative to the present climate for a climate scenario roughly equivalent to a doubling of atmospheric carbon dioxide

Figure 5.10 Hypothetical yield-temperature response curves for two crops (A and B) in the same region

Figure 5.11 Estimated crop yields under the GISS 2 × CO_2 scenario for present-day and for adjusted crop varieties and management conditions

Figure 5.12 Estimated maize yields in the USA under the GISS and GFDL 2 × CO_2 climates with and without the direct effects (DE) of CO_2: a) dryland and b) irrigated

Figure 5.13 Estimated wheat yields in the USA under the GISS and GFDL 2 × CO_2 climates with and without the direct effects (DE) of CO_2: a) dryland and b) irrigated

Figure 6.1 Estimated changes (%) in land use, by region, in the USA in response to changes in crop yields under the GISS and GFDL 2 × CO_2 climates

Figure 6.2 Estimated effects of climatic changes on agricultural production in Saskatchewan

Figure 6.3 Estimated effects of climatic variations on agricultural production in Hokkaido, Tohoku and all Japan

Figure 6.4 Estimated effects of climatic variations on agricultural production in Iceland

Figure 6.5 Estimated effects of climatic variations on agricultural production in northern, central and southern Finland

Figure 6.6 Estimated effects of climatic variations on agricultural production in the northern European USSR

Figure 7.1 Simulated agricultural effects of perturbed "climate" versus control runs to the year 2000 using the International Futures Simulation Model

Figure 7.2 Estimated changes in price of primary agricultural products to a range of yield reductions in the USA, the European Community and Canada

Figure 8.1 Simulated year-to-year variations in rice yield under observed (present-day) climate and GISS 2 × CO_2 climates for the period 1974–83 in Hokkaido (N. Japan)

Figure 8.2 Effects of adjustments to crop allocation in the Central Region (northern USSR) on agricultural receipts and production costs

TABLES

Table 1.1 Major net cereal importers, 1988

Table 1.2 Major net cereal exporters, 1988

Table 2.1 Estimates of changes in areal means of surface air temperature and precipitation over selected regions from pre-industrial times to 2030, assuming the IPCC Business-As-Usual Scenario

Table 2.2 Changing odds of a heatwave in selected cities if mean temperature increases by 3°F (1.7°C)

Table 4.1 Mean predicted growth and yield increases for various groupings of C_3 species for a doubling of atmospheric CO_2 concentration from 330 ppmv to 660 ppmv

Table 4.2 Response of spring-wheat yield (as percentages of the long term mean) to variations in air temperature (ΔT) and rainfall (ΔR) during the growing season (Cherdyn forest zone)

Table 4.3 Response of spring-wheat yield (as percentages of the long-term mean) to variations in air temperature (ΔT) and precipitation (ΔP) during the growing season (Palassovka, Volgograd region)

Table 7.1 Simulated global grain production in the year 2000 under a large warming scenario, as a percent of base level projections

Table 7.2 Global climate warming scenario used by Canadian AES study

Table 7.3 Changes in production opportunities estimated by Canadian AES study

Table 7.4 Assumed changes in yield (%) under an altered climate

Table 7.5 Change in welfare, by country or region, under three scenarios of climatic change

Table 7.6 Scenarios of food output under altered climate

ABBREVIATIONS AND ACRONYMS

AES	Atmospheric Environment Service (Canada)
CFCs	Chloroflurocarbons
CIAP	Climate Impact Assessment Program
ENSO	El Nino/Southern Oscillation
EPA	Environmental Protection Agency (U.S.)
ETS	Effective Temperature Sum
FAO	Food and Agriculture Organisation of the United Nations
GCM	general circulation model
GFDL	Geophysical Fluid Dynamics Laboratory
GHG	greenhouse gases
GISS	Goddard Institute for Space Studies
IFIAS	International Federation of Institutes of Advanced Study
IFS	International Futures Simulation
IIASA	International Institute for Applied Systems Analysis
IPCC	Intergovernmental Panel on Climate Change
ITCZ	Intertropical Convergence Zone
NCAR	National Centre for Atmospheric Research
NDU	National Defense University (U.S.)
OSU	Oregon State University
SWOPSIM	Static World Policy Simulation

UKMO	United Kingdom Meteorological Office
UNEP	United Nations Environment Programme
USDA	United States Department of Agriculture

PREFACE

This book has its immediate roots in an assessment by the Intergovernmental Panel on Climate Change (IPCC) of the potential impacts of climate change on agriculture. As lead author of that assessment I was aware of the many omissions that were required in our need to condense the results of more than 150 scientific papers by 50 scientists into less than 30 pages. Being an expansion of the IPCC assessment and in attempting to review much of what we currently know about climate change and agriculture, this book inevitably draws upon the work of many scientists – far too many to name individually. Some, however, deserve mention either because of their contribution to the subject or the inspiration they gave to others. There are for example the 200 or so who worked on the two 800-page IIASA/UNEP volumes published in 1988, particularly my colleagues at IIASA, Tim Carter and Nico Konijn. More recently I have leant heavily on researchers in the AIR Group at the University of Birmingham: Tim Carter, Tom Downing, Cynthia Parry, Julia Porter and John Wright. These and other colleagues have given unstintingly of their time and advice. Sue Pomlett keyboarded and corrected the text. Jean Dowling, Kevin Burkhill and Tim Grogan drew the figures. Cynthia Parry proofread and indexed the text. I thank them all, especially Cynthia, and hope that they can take some satisfaction from the result.

Martin Parry
Birmingham
June, 1990

1. THE SENSITIVITY OF AGRICULTURE TO CLIMATE

INTRODUCTION

In 1990 the Intergovernmental Panel on Climate Change (IPCC) completed its report on the greenhouse effect. The IPCC had been set up under the auspices of the World Meteorological Organisation and the United Nations Environment Programme, to examine how climate and sea level might change, what might be the impact of these changes and what could be the most appropriate response to them. IPCC Working Groups tackled each of these three tasks. Working Group II (Impacts) concluded that greenhouse gas-induced changes of climate would have an important effect on agriculture, with the most severe negative impacts probably occurring in regions of high present-day vulnerability that are least able to adjust technologically to such effects.[1] The purpose of this book is to consider, in more detail than could be covered within the confines of the IPCC report on agriculture, the reasoning behind this conclusion, its implications for global food security and the most appropriate courses of action.

GLOBAL WARMING

The best judgement of the IPCC is that, if emissions of greenhouse gases (GHG) continue to grow as currently projected (a so-called "Business-As-Usual" scenario), then global mean temperatures will increase by 0.2°C-0.4°C per decade over the next century.[2] There is a quite clear indication that a warming of the globe has occurred over the past century, amounting to 0.3°C-0.6°C. Much of this warming has been concentrated in two periods, between about 1920 and 1940 and since 1975; the six warmest years on record have all been in the 1980s. The size of this warming is broadly consistent with the predictions of climate models but, because of the natural variability of the Earth's climate, IPCC scientists are not yet able to say that they have detected the unequivocal "signal" of man-made climate change.[2]

The substantial uncertainties surrounding this issue are considered in the next chapter. For the present, let us consider in outline how the climate could change and, if it did, what would be the most likely consequences for world food supply.

Warming in high latitudes

There is relatively strong agreement that greenhouse gas-induced warming will be greater at higher latitudes.[3] This would reduce current temperature constraints on agriculture and probably increase productive potential, particularly in northern parts of North America, Europe and Asia. Soil and terrain constraints are, however, likely to limit the actual increase in agricultural output here, and it is probable that such gains in potential at high latitudes will do little to compensate for quite possibly substantial losses in potential in mid- and low latitudes.[4]

Poleward advance of monsoon rainfall

In a warmer world monsoon rains would be likely to penetrate further poleward, both in Africa and Asia, as result of an enhanced ocean–continent pressure gradient (itself the result of more rapid warming of the land than the ocean in the pre-monsoon season).[5] If this were to occur – and it should be emphasized that there remains much uncertainty here – then total rainfall could increase in currently drought-prone regions such as the Sahel and north-west India. It is possible, however, that the increase in rainfall would come largely in the form of more intensive rainstorms occurring over a shorter rainy period. If current levels of pre-monsoon rains, which are important for the germination of crops at the beginning of the growing season, were to diminish then growing seasons could be shortened and thus the potential for agriculture reduced. In addition, more intense rainfall could exacerbate problems of flooding and soil erosion.

Reduced crop-water availability

Probably the most important consequences of projected changes in climate for agriculture would stem from higher actual evapotranspiration, primarily as a result of higher temperatures of the air and land surface. Even in the tropics, where temperature increases are expected to be smaller than elsewhere, the increased rate of moisture loss from plants and soil would be considerable. It may be somewhat reduced by greater humidity and increased cloudiness during rainy seasons, but could be more pronounced in dry seasons.

Further details of these projected changes of climate, including the substantial uncertainties surrounding them, will be given in Chapter 2, and their implications for agriculture will be considered in subsequent chapters of this book. In one respect, however, the degree of vulnerability of agriculture to possible changes of climate is as much determined by the present-day vulnerability to weather as by the future patterns of climate change. This is considered in the next section.

THE MOST VULNERABLE REGIONS

In many regions of the world agricultural production is currently limited by climate, most of this limitation being in developing countries. Insufficient rainfall is the main climatic limit in these areas, curtailing the growing period available for crops. Overall, 63 per cent of the land area of developing countries is climatically suited to rainfed agriculture, but this endowment varies considerably between regions;[6] it amounts to as much as 85 per cent in South America and 84 per cent in south-east Asia, but is limited to 64 per cent in Central America and 53 per cent in Africa. The severest climatic limitations to agriculture are to be found in south-west Asia where 17 per cent is too mountainous and cool, and 65 per cent too dry, leaving only 18 per cent as potentially productive.[6]

The potential base for rainfed agriculture is therefore very limited in some regions and any further curtailment of potential due to changes of climate could severely strain their ability to feed local populations. Regions where climatic and soil resources are considered by FAO to be unable to meet the current needs of local populations are indicated in Figure 1.1. They occupy as much as 22 per cent of the global land area and contain 11 per cent of the world's population. They are mainly located in the cool and cold tropics (e.g. the Andean region, the Maghreb in North Africa, the mountain regions of south-west Asia), the Sahel and the Horn of Africa, the Indian subcontinent, and parts of mainland and insular south-east Asia. We shall show later that some of these regions could be the most at risk from possible future changes of climate.

Figure 1.1 Regions identified as critical zones in respect of ability to support current population. (*Source:* FAO, 1984).[6]

Critical with high inputs

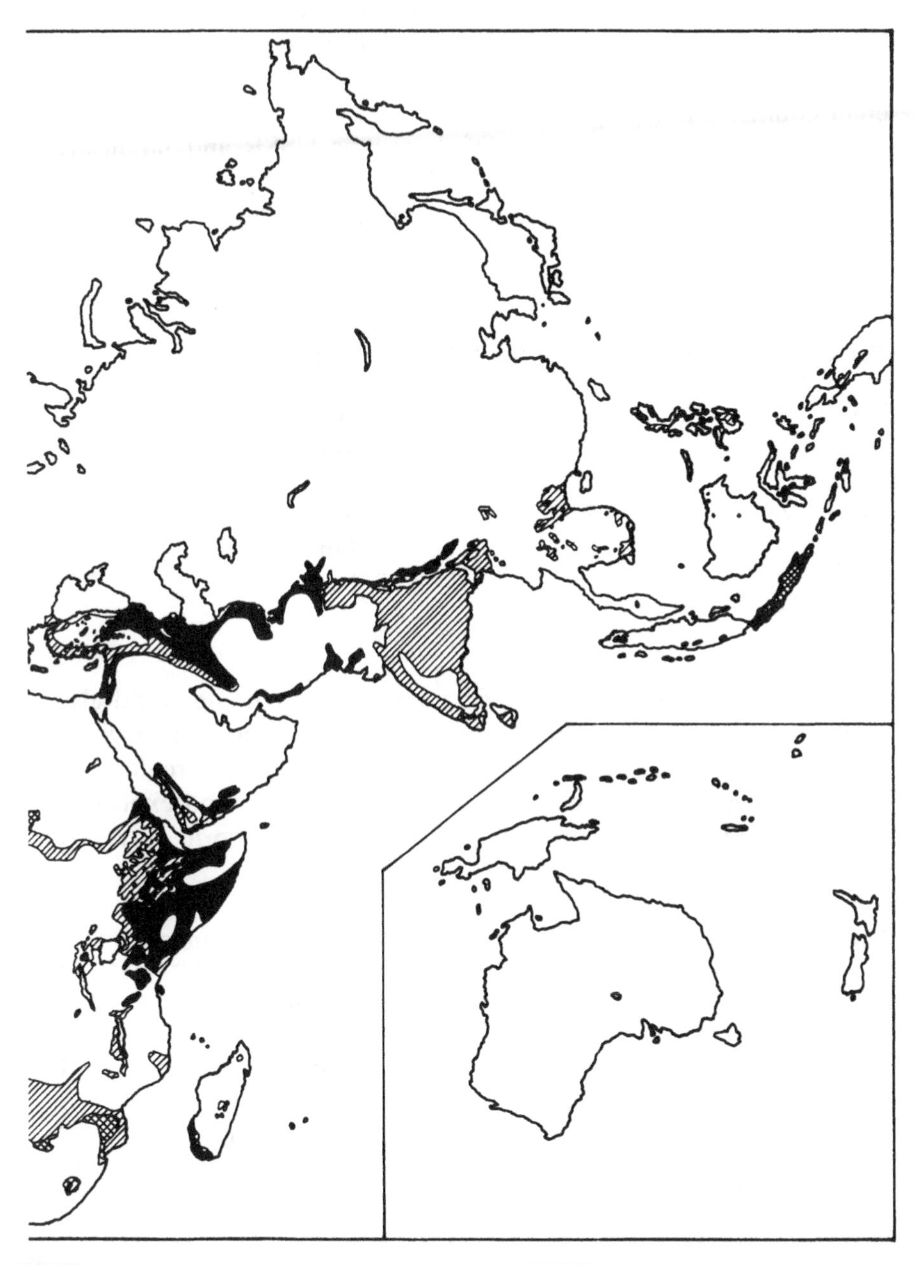

 Critical with intermediate inputs  Critical with low inputs

SENSITIVITY OF THE WORLD FOOD SYSTEM

Other countries that may become intrinsically vulnerable to impacts of climatic change are those that are currently major net food importers and thus whose national balance of payments can be substantially affected by world food prices. At present there are eight countries that import more than about five million tonnes of cereals each year (Table 1.1). Of these, the USSR and Japan are by far the major importers, but developing countries which are also substantial net importers may be the most vulnerable (e.g. Egypt, South Korea, Mexico).

Table 1.1: Major net cereal importers, 1988 (million tonnes)

USSR	34	Mexico	6
Japan	28	Iran	5
China	17	Italy	5
Korea, Rep.	9	Iraq	4
Egypt	8	Saudi Arabia	3

Source: FAO 1988.[7]

Substantially adverse or beneficial changes of climate could also markedly affect the amount of traded food and its price through its potential effects on output in the current food-exporting regions. A key role in the world food system is played by a few food-exporting countries, with only 21 out of 172 countries in the world currently

Table 1.2: Major net cereal exporters, 1988 (million tonnes)

USA	98	Thailand	6
France	27	Denmark	2
Canada	23	United Kingdom	1
Australia	15	South Africa	1
Argentina	10	New Zealand	–

Source: FAO 1988.[7]

being net cereal exporters (Table 1.2). In 1988 three countries accounted for 80 per cent of all traded cereals (USA, Canada and

France), with over one-half exported by the USA alone (including well over a half of the world's traded maize and three-quarters of its traded soya beans).[7]

These major food-producing countries are doubly important to the world food system because of their key role as holders of large food stocks. In 1988/89 the USA, Canada and the European Community held almost one-third of the world's stock of wheat and coarse grains (maize, sorghum and barley). The USA alone held 17 per cent of wheat and 47 per cent of maize.[8]

In addition, it should be noted how sensitive world food security remains to variations of weather. In a good year world food production now exceeds demand by about 20 per cent, but a relatively short run of poor years can eliminate this excess supply. To illustrate, in 1987/88 world wheat and coarse grain stocks stood at 353 million tonnes (mt), the equivalent of about 78 days supply, but fell to 248 mt (54 days) in 1988/89 largely as a result of the 1988 drought and heatwave in the US Corn Belt and Great Plains.[8] Stocks of wheat in the USA alone fell from 49 million tonnes in 1986/87 to 34 in 1987/88 and to 17 in 1988/89. What happens to food stocks in North America thus also affects world food stocks and prices.

These "breadbasket" countries are critical to the world food system, being regions where relatively small changes in food production due to changes of climate could have a severe effect on the quantity, price and type of food products bought and sold on the world food market.

MARGINAL FARMERS

Whether or not they are located in resource-poor countries, there is also a strong indication that marginal agriculture and marginal farmers may be most vulnerable both to short-term variations of weather and longer-term changes of climate.[9,10] This marginality can be construed in a number of ways – spatial or economic or social. Agriculture may be marginal in a spatial sense when types of farming are practised at or near the edge of their appropriate climatic region (for example, the northern limit of sufficiently reliable rainfall for cereal farming in the Sahel; or the southern limit of sufficiently high temperatures for coffee production in southern Brazil). Relatively small changes of climate in these areas could substantially alter the potential for agriculture thus creating a mismatch between existing farming systems and prevailing climatic resources for agriculture.

Economic marginality, that is, where returns to farming barely exceed costs, may frequently characterize the regions described above, but it may also stem from inadequate infrastructure, under-capitalization or other factors. Whatever the cause, marginal farmers with low levels of financing are likely to have fewer resources to deploy in adapting to climatic change.

Finally, there are some rural groups that may be described as socially marginalized, that is, where they have become isolated from their indigenous resources base and forced into marginal economies that contain fewer adaptive mechanisms for survival.[11]

In each of these instances of marginality there may thus be a proneness to impact from changes of climate simply because of the limited resources available for the adaptation needed either to benefit fully from positive effects of changes of climate or to mitigate successfully the negative effects.

THE STRUCTURE OF THIS BOOK

The possible changes of climate due to greenhouse-gas forcing are outlined in the next chapter. There then follows some discussion of the methods by which we can begin to assess the potential effects of such changes on agriculture (Chapter 3). The first task is to estimate impacts of a biophysical nature on plant and animal growth, on pests and diseases and on the soil (Chapter 4). This enables some assessment to be made of consequent changes in overall agricultural potential, such as in the geographical limits to different types of farming and in levels of agricultural yield (Chapter 5). It is then possible to consider the possible consequences for food production (Chapter 6) and for global food security (Chapter 7). Chapter 8 evaluates a range of possible adaptations by agriculture both to mitigate those effects of climatic changes that are potentially negative and to enhance those that offer potential benefits. A final chapter of conclusions follows.

2. POSSIBLE CHANGES OF CLIMATE

THE GREENHOUSE EFFECT

Solar radiation received at the Earth's surface provides the energy that fuels all life on Earth. Being a very hot star (6000 K), the Sun produces shortwave radiation, most of which penetrates the Earth's atmosphere and reaches its surface but some of which is scattered or absorbed during passage through the atmosphere. The Earth itself emits radiation but, being much cooler than the Sun, does so at longer wavelengths. If the Earth had no atmosphere, and was completely transparent to this outgoing longwave radiation, then its temperature would probably be only about −18°C (about the same temperature as the planet Mars which has no atmosphere). Fortunately the Earth's atmosphere contains a number of gases that tend to absorb, and thus block, part of the outgoing longwave radiation while allowing passage of much of the incoming shortwave radiation. This is enough to raise global surface temperatures by some 33°C to a much more comfortable average of 15°C. The atmosphere thus acts as a warming envelope in much the same way as the glass of a greenhouse, and the "greenhouse gases" of the atmosphere (largely carbon dioxide, water vapour and other trace gases) are thus crucial to life on this planet.

But increases in concentration of greenhouse gases can upset the Earth's radiation balance, with more of the longwave radiation being absorbed in the lower atmosphere and some of this being re-emitted back to the Earth's surface. This enhanced Greenhouse Effect could cause increases in the Earth's surface temperature great enough to threaten many of its life forms.

GREENHOUSE GASES

The most important greenhouse gas, in terms of recent changes in radiative forcing, is carbon dioxide (CO_2), followed by chlorofluorocarbons (CFCs), methane (CH_4) and nitrous oxide (N_2O).

Atmospheric concentrations of CO_2 have increased by about 25 per cent over the past two centuries, from about 280 parts per million by volume (ppmv) in about 1850 to 353 ppmv (1989), largely as a result of the burning of fossil fuels but also due to clearance of forests for agriculture. At present CO_2 concentrations are growing at the rate of about 0.5 per cent per year and are estimated to account for 57 per cent of radiative forcing.[1] Methane, which accounts for about 14 per cent of radiative forcing, has more than doubled in concentration in the atmosphere since the pre-industrial period, mainly due to increases in the extent of rice paddies and in numbers of ruminant livestock (methane is generated by the anaerobic decomposition of organic matter in the mud of paddy fields and in the stomachs of cattle and sheep).

Nitrous oxide, which accounts for about 5 per cent of radiative forcing has increased from about 288 parts per billion by volume (ppbv) to 310 ppbv over the last 250 years, and is increasing in atmospheric concentration at an annual rate of about 0.33 per cent, largely as a result of increased use of nitrogen fertilizers.[1]

CFC concentrations, although extremely small, are very powerful greenhouse gases and account for over 20 per cent of recent changes in radiative forcing. Until agreement on control of CFC use was achieved through the Montreal Protocol in 1989, atmospheric concentrations of the important CFCs were growing at an annual rate of 4.0%, largely as a result of the expansion of their use as propellants, solvents and blowing agents for foam packaging. With the Montreal Protocol the contribution of CFCs to radiative forcing should fall to about 5 per cent over the next 50 years.[1]

Agriculture itself is an important source of greenhouse gases. It contributes about 35 per cent of all current methane emissions, and a large but at present unspecified amount of all N_2O emissions.

Allowing for the different levels of radiative forcing due to these different gases, we may conclude that agriculture contributes about 15 per cent of current greenhouse-gas forcing.[2]

PROJECTED CHANGES IN GREENHOUSE GASES

It is extremely difficult to predict future changes in concentrations of greenhouse gases because of the large uncertainties regarding future patterns of energy production and land use. The projected range given in Figure 2.1 combines the various greenhouse gases in terms of their radiative equivalence to CO_2 and indicates this from 1985 to 2100. Under the Business-As-Usual scenario the date at which a doubling of the equivalent CO_2 above pre-industrial levels is reached (i.e. 560 ppmv) is about 2025. Under the low scenario (D) this will occur about 2070.

Figure 2.1 Radiative forcing, also expressed as equivalent CO_2 concentrations, resulting from four emissions scenarios selected by the IPCC. The Business-As-Usual scenario assumes few controls on GHG emissions or deforestation. Scenario B assumes stringent controls on carbon monoxide, CFCs and deforestation. Under Scenario C a shift towards renewables and nuclear energy takes place in 2050–3000. Under Scenario D a shift to renewables and nuclear occurs before 2050. (*Source:* IPCC, 1990).[1]

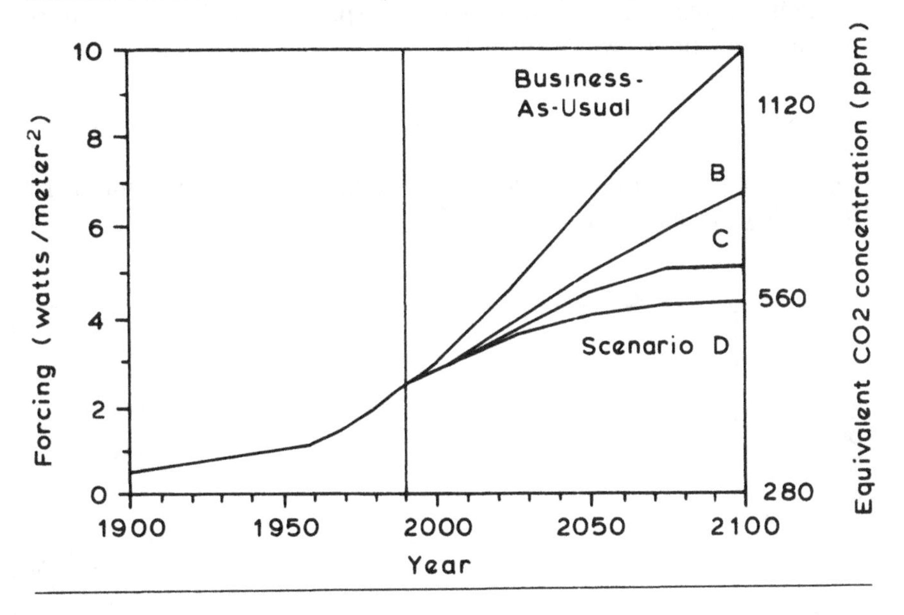

PROJECTED CHANGES IN CLIMATE

Several teams of climatologists have attempted to estimate the likely effect of the projected changes in greenhouse gas concentrations on the Earth's climate. The complex climate models they use ("general circulation models" or GCMs) are mathematical representations of the atmospheric, oceanic and terrestrial processes that are known to occur and that can be formulated in terms of equations capable of being solved by computers.

Six GCMs have been used in greenhouse-gas experiments. In most cases the models have been run assuming a "control" level of CO_2 of around 300 ppmv that approximates the pre-industrial level of concentration. The model is then run again with doubled CO_2 levels until the simulated climate reaches an equilibrium condition. The difference between the $1 \times CO_2$ and the $2 \times CO_2$ climates is taken as an indication of the climate response to an equivalent doubling of CO_2.

Although there are important differences between the GCMs and the results of their $2 \times CO_2$ experiments, there is some agreement that global mean surface temperatures should rise between 2°C and 4°C, that warming will be greater at higher than at lower latitudes, that warming will be greater in the winter half of the year than in the summer and that, on average, precipitation will increase due to higher rates of evaporation and transpiration and, thus, a more humid atmosphere will prevail.[1]

Figure 2.2 Predictions of global mean temperature change under the four IPCC scenarios. (*Source:* IPCC, 1990).[1]

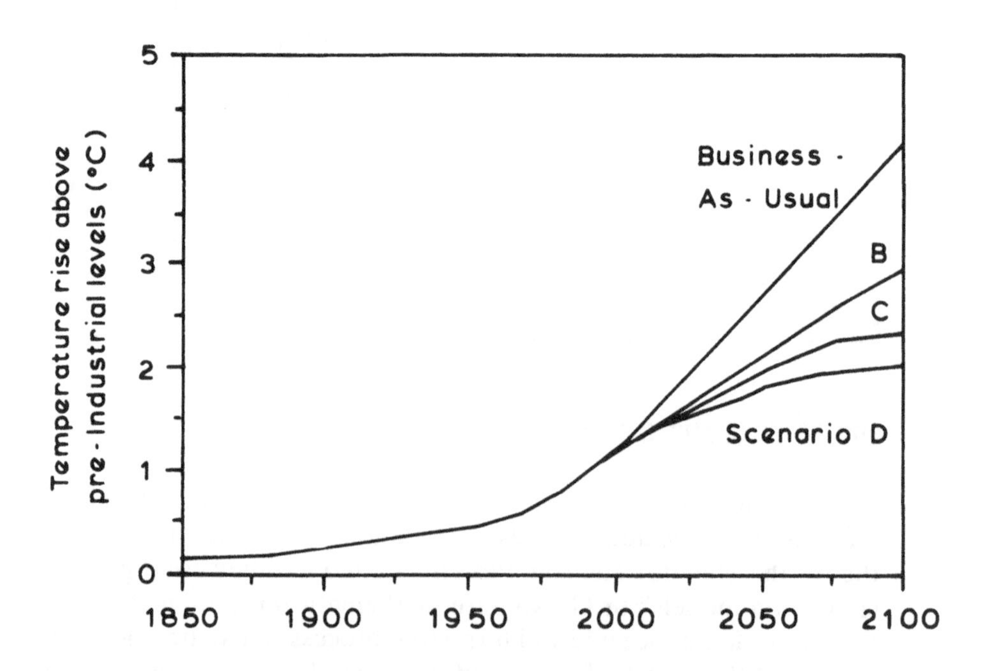

Figure 2.3 Model average of the 2 × CO_2 minus 1 × CO_2 change in surface air temperature (T). Results are a) December, January and February and b) June, July and August, and are computed using equilibrium response data from five GCMs (GFDL, OSU, GISS, NCAR and UKMO). The contour interval is 1°C. Shading denotes areas where T is greater than or equal to 5°C. (*Source:* Santer, *et al.* 1990).[3]

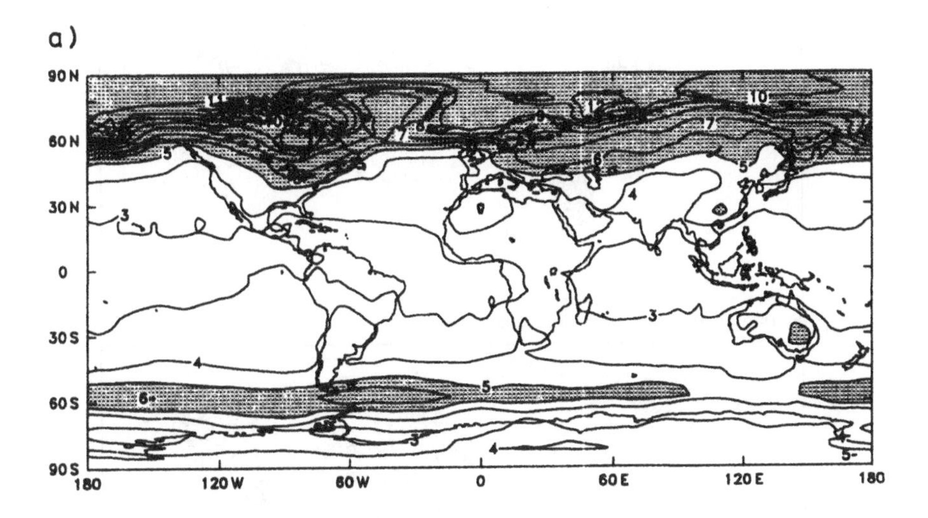

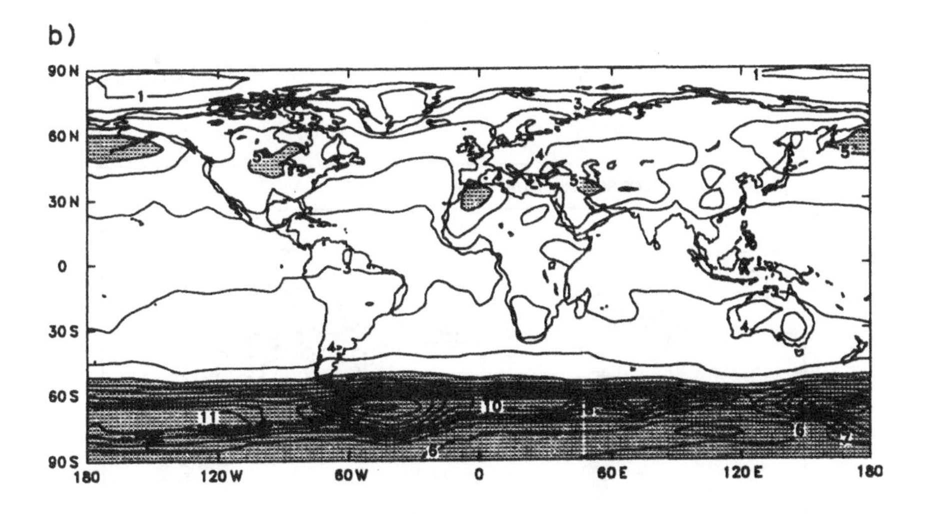

Figure 2.4 Changes in soil moisture due to doubling CO_2 as simulated by three high resolution models: (a) and (d) Canadian Climate Centre model; (b) and (e) Geophysical Fluid Dynamics Laboratory; (c) and (f) United Kingdom Meteorological Office. Contours every 2 cm, areas of decrease lightly stippled. Note that the Canadian

(a) DJF 2 X CO2 - 1 X CO2 SOIL MOISTURE: CCC

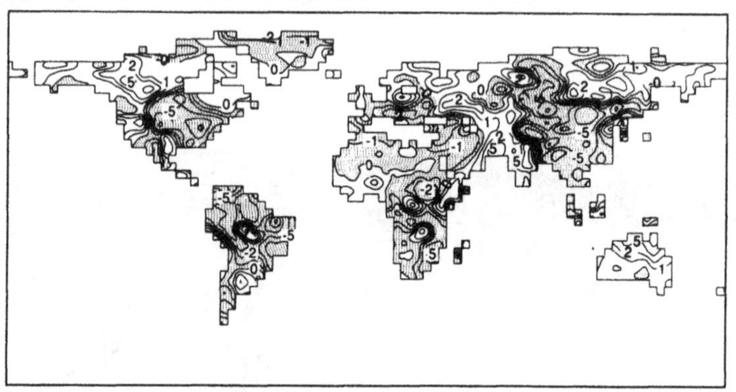

(b) DJF 2 X CO2 - 1 X CO2 SOIL MOISTURE: GFHI

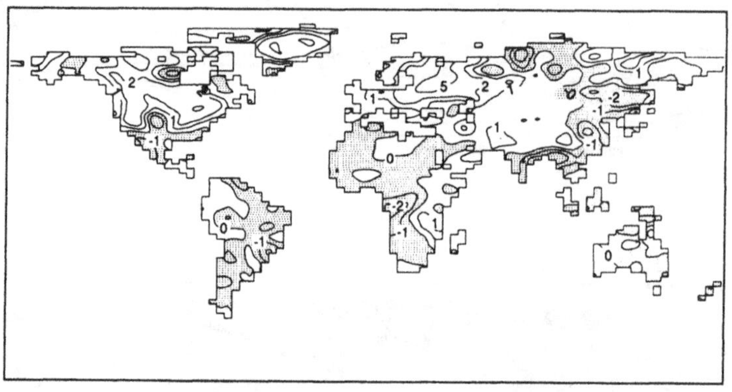

(c) DJF 2 X CO2 - 1 X CO2 SOIL MOISTURE: UKHI

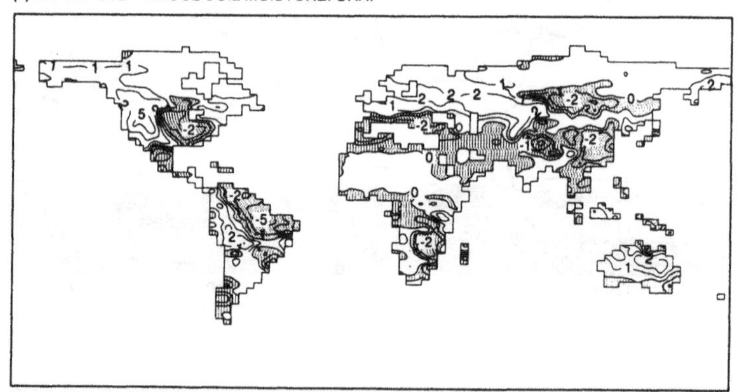

Climate Centre model (a, d) has a geographically variable soil capacity whereas the other two models have the same soil capacity everywhere. (a), (b), (c) December January and February, (d) (e) (f) June, July and August. (*Source:* Mitchell, *et al.* 1990).[4]

(d) JJA 2 X CO_2 - 1 X CO_2 SOIL MOISTURE: CCC

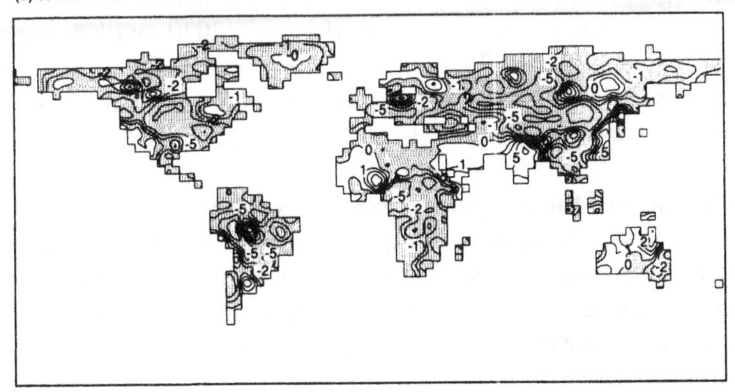

(e) JJA\2 X CO_2 - 1 X CO_2 SOIL MOISTURE: GFHI

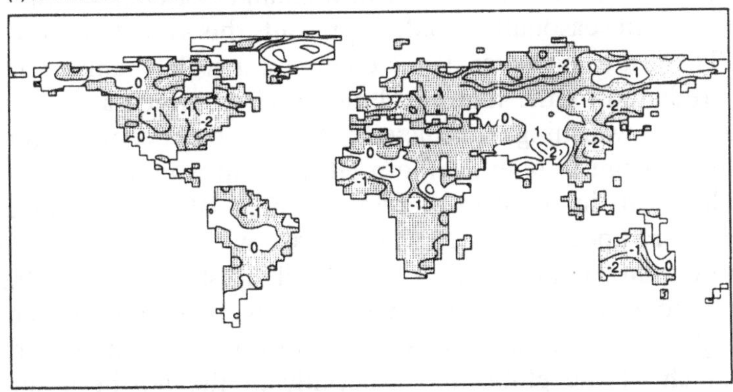

(f) JJA 2 X CO_2 - 1 X CO_2 SOIL MOISTURE: UKHI

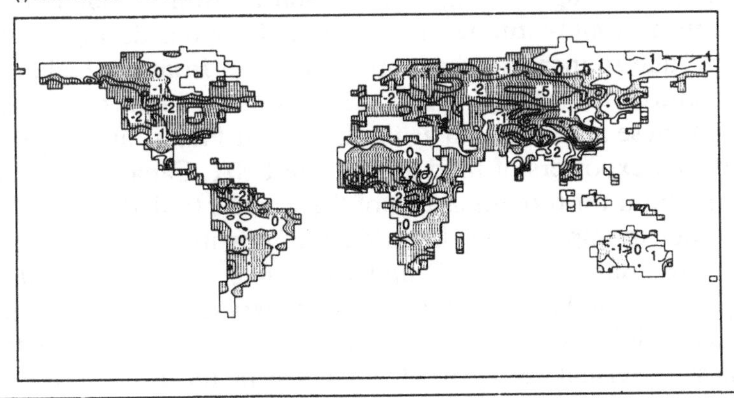

Most of the GCM greenhouse-gas experiments have simply doubled the CO_2 concentration in the atmosphere. This is unrealistic since CO_2 concentrations are increasing gradually over time and it also fails to indicate the possible change in climate over the next 20 or 30 years which is a more realistic time frame for consideration of likely impacts and responses. Under the IPCC Business-As-Usual scenario global mean temperatures are estimated to rise about 2°C above pre-industrial levels by the year 2030 (with a range of uncertainty of 1.4°C to 2.9°C). This is equivalent to a rise of 1.1°C above present-day temperatures. By 2090 temperatures would be 4°C above pre-industrial levels, equivalent to a rise of 3.3°C from today (Figure 2.2).

An average of the temperature changes estimated by five $2 \times CO_2$ GCM experiments is given in Figure 2.3. There is broad agreement between the projections for larger increases at higher than at lower latitudes and larger increases in winters than in summers. As regards precipitation, however, there is less agreement and an average precipitation scenario is not meaningful. To determine likely effects on agriculture we need to know how changes of climate will occur regionally and seasonally – and, at present, this knowledge does not exist. There are, however, some continental-scale changes, consistently predicted by the high resolution GCMs, which are plausible. For example, warming is predicted to be 50-100 per cent greater than the global mean in high northern latitudes in winter. Precipitation is predicted to increase in middle and high latitude continents in winter.[1, 4] Evaporation rates will increase with higher temperature (by more than 5 per cent per °C), and soil moisture can therefore be expected to decrease in many areas where the precipitation increases are small. As a consequence, soil moisture is expected to decrease over much of the globe and especially in the northern summer months (June-August) (Figure 2.4). Soil moisture reductions may be greatest in mid-latitude midcontinental regions during northern summer due to increased evapotransporation from the higher land temperatures in summer.[4] This could well be important because some of these regions, such as the US Great Plains, are major producers and exporters of food today. The USA, Canada and France alone account for three-quarters of the world's traded cereals.[5]

In order to improve our knowledge of possible regional patterns of climate change the IPCC undertook five regional case studies using results of high resolution GCM experiments. The studies assumed a global mean warming of 1.8°C at 2030 (consistent with the Business-As-Usual scenario). Confidence in the estimates is low, especially for the changes in precipitation and soil moisture, but represent the current best estimate. These uncertainties imply a range from approximately 70 per cent to 145 per cent of the estimates below. The following summary and the accompanying details given in Table 2.1 are taken from the IPCC report:[4]

Central North America (35-50°N, 85-105°W)
The warming varies from 2 to 4°C in winter and 2 to 3°C in summer. Precipitation increases range from 0 to 15% in winter whereas there are deceases of 5 to 10% in summer. Soil moisture decreases in summer by 15 to 20%.

South East Asia (5-30°N, 70-105°E)
The warming varies from 1 to 2°C throughout the year. Precipitation changes little in winter and generally increases throughout the region by 5 to 15% in summer. Summer soil moisture increases by 5 to 10%.

Sahel (10-20°N, 20° W-40°E)
The warming ranges from 1 to 3°C. Area mean precipitation increases and area mean soil moisture decreases marginally in summer. However there are areas of both increase and decrease in both parameters throughout the region which differ from model to model.

Southern Europe (35-50°N, 10° W-45°E)
The warming is about 2°C in winter and varies from 2 to 3°C in summer. There is some indication of increased precipitation in winter, but summer precipitation decreases by 5 to 15%, and summer soil moisture by 15 to 25%.

Australia (10-45°S, 110-155°E)
The warming ranges from 1 to 2°C in summer and is about 2°C in winter. Summer precipitation increases by around 10%, but the models do not produce consistent estimates of the changes in soil moisture. The area averages hide large variations at the sub-continental level.

Many of the differences in these results can be attributed to differences in model resolution, neglect or otherwise of ocean heat transport, and differences in the number of physical processes included and the way they are represented.

Table 2.1: Estimates of changes in areal means of surface air temperature and precipitation over selected regions, from pre-industrial times to 2030, assuming the IPCC Business-As-Usual Scenario. These are based on three high resolution equilibrium studies which are considered to give the most reliable regional patterns, but scaling the simulated values to correspond to a global mean warming of 1.8°C, the warming at 2030 assuming the IPCC "best guess" sensitivity of 2.5°C and allowing for the thermal inertia of the oceans. The range of values arises from the use of three different models. The range of uncertainty in global mean sensitivity (1.5 to 4.5°C) implies a range of approximately 70% to 145% of the values given below. *Confidence in these estimates is low*, particularly for precipitation and soil moisture. Note that there are considerable variations in the changes within some of these regions. The numbers 1, 2 and 3 in the third column refer to three models: 1, Canadian Climate Centre (CCC); 2, Geophysical Fluid Dynamics Laboratory (GFDL); 3, United Kingdom Meteorological Office (UKMO). *Source:* Mitchell, *et al.* (1990).[4]

Region		*Temperature (°C)*		*Precipitation (%)*		*Soil moisture (%)*	
		DJF	*JJA*	*DJF*	*JJA*	*DJF*	*JJA*
1 Central N America	1	4	2	0	−5	−10	−15
(35–50N,	2	2	2	15	−5	15	−15
80–105W)	3	4	3	10	−10	−10	−20
2 South East Asia	1	1	1	−5	5	0	5
(5–30N,	2	2	1	0	10	−5	10
70–105E)	3	2	2	−15	15	0	5
3 Sahel	1	2	3	−10	5	0	−5
(10–20N,	2	2	1	−5	5	5	0
20W–40E)	3	1	2	0	0	10	−10
4 Southern Europe	1	2	2	5	−15	0	−15
(35–50N,	2	2	2	10	−5	5	−15
10W–45E)	3	2	3	0	−15	−5	−25
5 Australia	1	1	2	15	0	45	5
(12–45S,	2	2	2	5	0	−5	−10
110–155E)	3	2	2	10	0	5	0

In almost all regions of the world, in both the core food exporting regions of today and in areas that are not self-sufficient in food, changes in crop-water availability are likely to be the most important for agriculture. While we cannot be certain about the regional pattern of soil moisture changes that may occur, there are regions in the world where GCM predictions are in some agreement. The following is a summary of those regions where 2 × CO_2 experiments by three GCMs (GFDL, GISS and NCAR) and the three high-resolution models shown in Figure 2.4 all project decreases in soil moisture.[6, 7] It should be emphasized that coincidence of results for these regions is not statistically significant and that the evidence available at present is extremely weak. Moreover, the significance of decreases in soil water will vary considerably from region to region according to whether they occur during the growing or non-growing season.

i) Decreases of soil water in December, January and February:
   Africa: North-east Africa, southern Africa
   Asia: western Arabian Peninsula; Southeast Asia
   Australasia: eastern Australia
   N. America: southern USA
   S. America: Argentine pampas
ii) Decreases in soil moisture in June, July and August:
   Africa: north Africa; west Africa
   Europe: parts of western Europe
   Asia: north and central China; parts of Soviet central Asia and Siberia
   N. America: southern USA and Central America
   S. America: eastern Brazil
   Australasia: western Australia

The regions identified above are those where crop water availability may decrease in an equilibrium climate for an equivalent doubling of atmospheric CO_2. But different latitudes could approach equilibrium at different rates of change in temperature and rainfall because they include different amounts of land, which warms up faster than ocean. Thus the time-evolving patterns of soil moisture change could vary significantly from the equilibrium simulation. Moreover the effects of climatic changes on agriculture are likely to be greater during periods of rapid change, before equilibrium is reached and before farming systems have had time to adapt to their altered environment.

Despite these caveats, however, we may conclude for the present that these are the regions where there are preliminary indications of potentially significant reductions in agricultural potential due to decreases in crop water availability. A comparison with Figure 1.1 indicates that several of these regions prone to greenhouse gas-induced drying are also those that are barely able to support existing populations from their own resource base. The regions which are particularly at risk (because they are both vulnerable now and may face increased drought in the future) include:
Africa: north Africa; north-east Africa; southern Africa
Asia: western Arabia; southeast Asia
N. America: Central America
S. America: eastern Brazil

CHANGES IN THE VARIABILITY OF CLIMATE

As important for agriculture as possible changes in mean climate may be changes in the variability of climate, particularly in the frequency of extreme weather events (such as severe storms, heat waves and damaging frosts) that today exact a major toll on food output. Few farmers, for example, plan their activities on their expectation of the average return to farming. They generally gamble on good years and insure against bad ones.[8, 9] Any changes in the frequency of good and bad years can thus have a major effect on profitability of agriculture. Moreover we know that the relationship between mean climate and the frequency of extreme events can be strongly non-linear and that quite small changes in the mean can significantly alter the frequency of extremes.[9, 10]

Unfortunately, little is known at present about the likely changes in measures of climatic variability due to greenhouse gas forcing. Much more work on this issue is needed before we can fully evaluate the effect of greenhouse gas forcing on the effects of extreme weather on agriculture. However, even if we assume that the same distribution of extremes around the mean is maintained, changes in mean climate can be expected to increase markedly the frequency of extreme events such as hot days likely to cause heat stress for crops (Table 2.2). The potential effects of this on agriculture are considered in Chapter 4. At present we do not know whether tropical or mid-latitude storms will increase or decrease.[1]

Table 2.2: Changing odds of a heatwave in selected cities if mean temperature increases by 3°F (1.7°C)

City	*Maximum temperature*	*Odds now (%)*	*Odds if +3°F (%)*
Washington DC	95°F	17	47
Des Moines, Iowa	95°F	6	21
Dallas, Texas	100°F	38	68

Note: A heatwave is five or more days in a row at or above the maximum temperature indicated in the first column.
Source: Mearns *et al.* (1984).[11]

DETECTING THE GREENHOUSE EFFECT

On the basis of our understanding of the likely effect of greenhouse gas concentrations on climate, and knowing that concentrations have increased from 280 ppmv in about the mid-eighteenth century to about 353 ppmv at present (1989), we may estimate that the world should have warmed by between 0.5°C and 1.2°C by 1988.[1] In fact, the world has warmed by O.3°C-0.6°C since the late nineteenth century (Figure 2.5). A warming trend is evident up to about 1940, with relatively little change between the 1940s and 1970s. However, from about the mid-1970s warming has resumed at a quite rapid rate, with the six warmest years in the past century occurring in the 1980s, and the years 1987 and 1988 being the warmest on record.

The observed warming of the Earth is thus broadly consistent with the model predictions. The problem, however, is that climate varies naturally on a wide range of timescales as a result of many factors such as volcanic eruptions, changes in ocean circulation and solar variations. Until the "signal" of the greenhouse effect has risen above the "noise" of natural climatic variability we cannot be sure that the greenhouse effect has been detected. But if we wait until we are sure, the amount of warming to which we are committed will be greater and quite probably more costly to adapt to.

The greenhouse issue thus embodies some large uncertainties together with potentially enormous impacts. At present it is by no means clear whether we have sufficient evidence to determine the most appropriate courses of action. Would it, for example, be cheaper in the long run to mitigate the greenhouse effect by cutting carbon emissions rather than adapting to the unmitigated climate

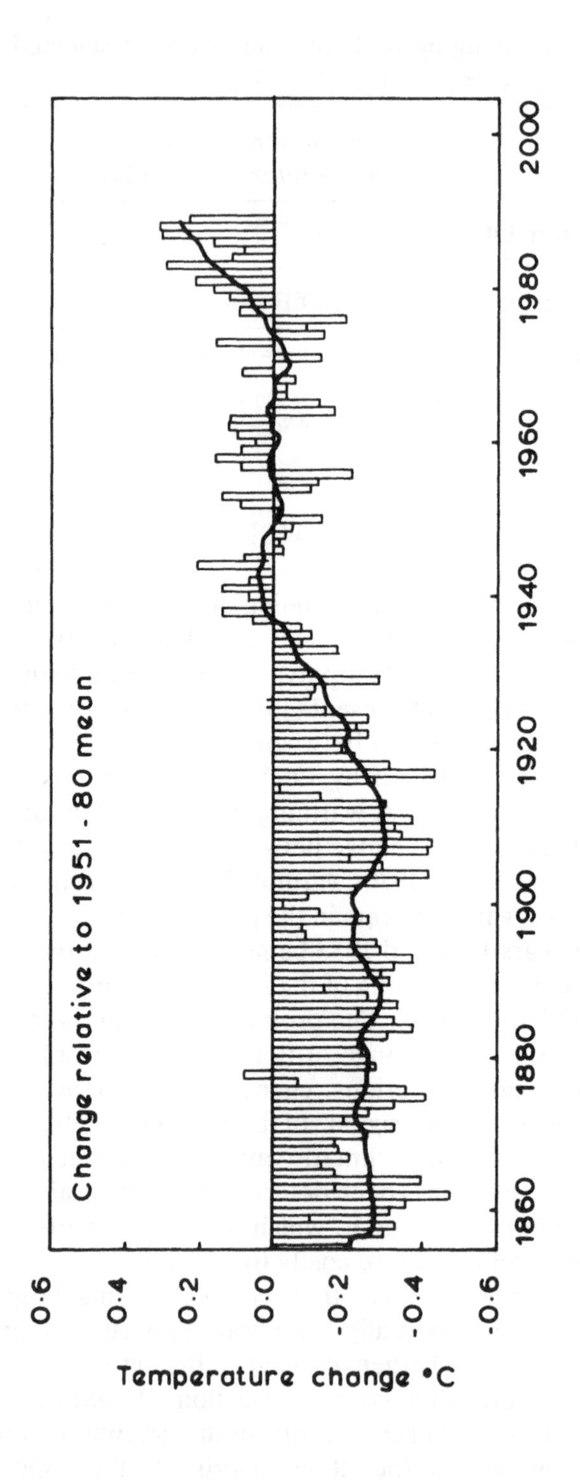

Figure 2.5 Global mean combined land-air and sea-surface temperatures, 1861–1989, relative to the average for 1951–80. (*Source:* IPPC, 1990).[1]

change? And what combination of mitigation and adaptation would make most sense? An effective answer to these questions requires a knowledge of the likely costs and benefits of different courses of action; these, in turn, require a knowledge of potential impacts and their costs, under scenarios both of high and low mitigation. The methods of assessing such impacts are discussed in the next chapter.

3. METHODS OF ASSESSING IMPACTS OF CLIMATIC CHANGE

THE DEVELOPMENT OF METHOD

The application of techniques of impact assessment to the field of climatic change is quite new. It began in the mid-1970s when concern first arose over atmospheric ozone depletion, and was first applied to the issue of greenhouse gas-induced climate change in the early 1980s. Since then there have been some substantial advances in method, from an initial focus on the one-way impact of climatic changes on human activity to, more recently, a greater emphasis on the two-way interactions between climatic change and human activity.

The early emphasis on impacts

The impact approach is based on the assumption of a direct cause and effect relationship between a climatic event (such as a short-term decrease in rainfall) and a response within the system under study (such as a decrease in crop yields, or migration of wildlife). A schema of this is given in Figure 3.1a. In impact assessment the system under study is sometimes referred to as an "exposure unit".[1]

Such an approach was characteristic of several assessments in the 1970s that used regression models to infer statistical relationships between climate change and potential impacts. An example is the Climate Impact Assessment Program (CIAP) funded by the US Department of Transportation in the early 1970s to investigate possible impacts of ozone depletion, including those that might be caused by supersonic commercial aircraft.[2] The tendency was to rely on extrapolations over the long-term of statistical relationships developed over a relatively short run of years, with little understanding of the processes that relate the cause to the supposed effects.

The impact approach also characterized an extensive effort in the late 1970s to estimate the possible effects of long-term climatic change on crop yields and agricultural production. This was conducted by the National Defense University in Washington, DC in response partly to the calamitous effects of persistent drought in the Sahel in the mid-1970s and partly to the concern that had emerged over the possible effects on the balance of power between the USA and USSR of severe drought in the major grain-growing regions of the superpowers.[3] Once again, the emphasis was on the search for a relatively simple connection between a potential climatic change and its likely effect on an exposure unit (in this case, food production). With the benefit of hindsight we now know that so many intervening factors can operate that it may be quite misleading to treat the three study elements (climatic change – exposure unit – impact) in isolation from their milieu. For example, many of the effects of reduced rainfall may not only be felt directly by plants and animals through a decrease in crop-water availability, but also through changes in soil structure and nutrient status due to rates of soil erosion being altered under drier conditions.

The subsequent focus on interactions

Since about 1980 there has been increasing emphasis on attempting to understand the interactions that might relate climatic change to

Figure 3.1 Schema of a) impact and b) interaction approaches in climate impact assessment. (*Source:* Parry & Carter, 1988).[4]

a) Impact approach

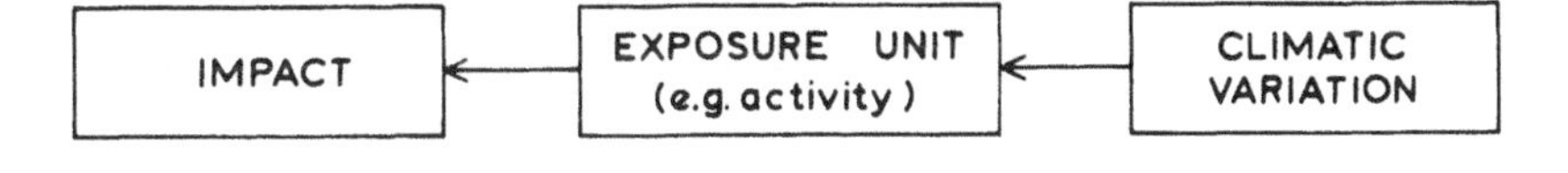

b) Interaction approach

its potential response by assuming that a climatic event is merely one of many processes (both biophysical and socio-economic) that may affect the exposure unit. A schema of this is given in Figure 3.1b. This approach was pioneered by the project on Drought and Man of the International Federation of Institutes of Advanced Study (IFIAS)[5] which investigated the role of drought in the deteriorating social and economic conditions in the Sahel in the 1970s. Its overall conclusion was that drought had merely triggered a crisis that had, in fact, been pre-conditioned by economic and political developments over the previous decades, particularly the encouragement of intensive forms of agriculture that were much more sensitive to short-term decreases in rainfall than the traditional forms of nomadic herding.

Figure 3.2 Schema of interaction approach to climate impact assessment, with ordered interactions. (*Source:* Parry & Carter, 1988).[4]

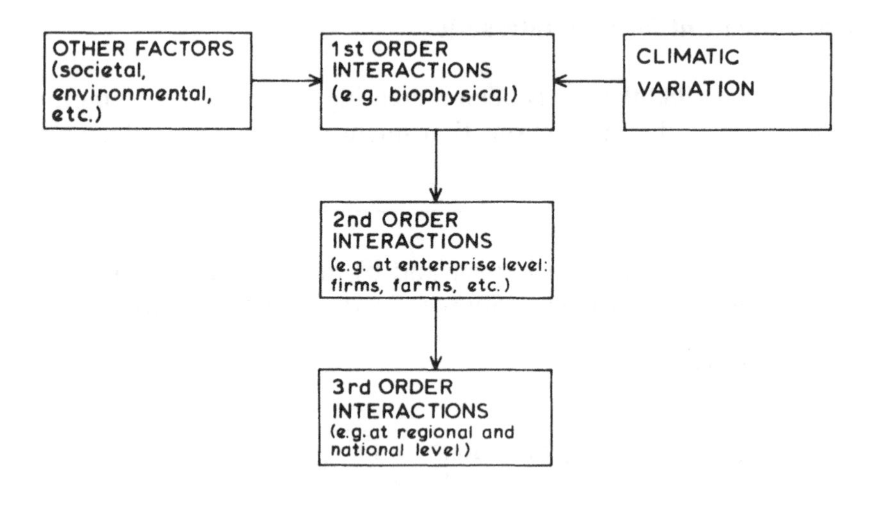

The interaction approach achieved greater realism by considering the "cascade" interactions that can occur between different elements in the exposure unit, in particular from the first-order biophysical level, through a second-order level characterized by units of enterprise to third-order interactions at the regional, national and international level (Figure 3.2). In agriculture first-order effects include those responses at the plant and animal level; second-order effects those at the farm level (e.g. farm-level production, employment, profitability, etc.); third-order effects those such as regional or national levels of output, trade, prices, etc.

One of the first studies to consider different orders of interaction in the response of agriculture to climatic change was a European Commission project on socio-economic effects of CO_2-induced climatic changes, which used outputs from GCMs to generate information on possible changes in temperature and precipitation that could be used as inputs to models of runoff and biomass production. However, economic interactions at the second- and third-order levels were not considered.[6]

Figure 3.3 A hierarchy of models for integrated climate impact assessments. (*Source:* Parry & Carter, 1988).[4]

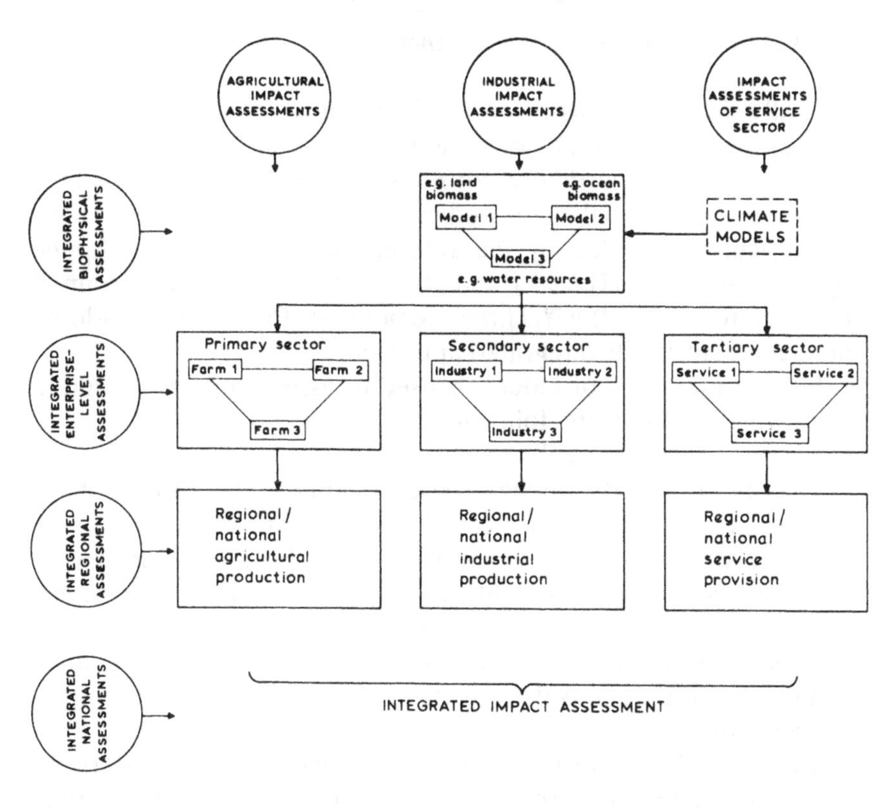

Integrated impact assessment

A fully integrated impact assessment assumes that study is made not only of interactions between different orders, but also of interactions within the same order, both within individual sectors (such as between different farming systems) and between different sectors (such as between the reinforcing or countervailing effects of changes in climate on agriculture, forestry, water resources, etc.) and the feedback effects operating between them. A schema is given in Figure 3.3. This form of fully integrated assessment has been proposed in some studies of a purely methodological kind, such as that by the Battelle Institute,[7] but it will not be possible to implement it until the full complement of systems models, for all sectors at each level, has been developed and satisfactorily tested in an interactive manner. For the present we must remain content with partially integrated impact assessments.

CURRENT STATE OF THE ART

Partially integrated impact assessments

Partially integrated assessments have now been completed for agriculture for a number of regions. A schema of one is given in Figure 3.4. The particular form illustrated here was developed for a set of regional case studies of impacts of climatic change on agriculture funded by the International Institute for Applied Systems Analysis (IIASA) and the United Nations Environment Programme (UNEP).[8, 9] Three elements in the approach adopted here characterize its integrated approach.

First, a hierarchy of models is used to assess impacts on agriculture, which includes the following:

- Models of climatic change (based mainly on GCM outputs but also on analysis of the observed climatic record)
- Models of first-order relationships (i.e. those between climatic variables such as temperature, rainfall and plant growth, crop yield, rangeland carrying capacity)
- Models of second-order, largely economic, relationships at the farm level, that consider effects of yield on production, employment, profitability
- Models of higher-order, also largely economic, relationships that consider effects on production, on regional and national output,

Figure 3.4 Schema of the IIASA/UNEP project's approach: an interactive approach to climate impact assessment with ordered interactions, interactions at each level, and some social and physical feedbacks. (*Source:* Parry & Carter, 1988).[4]

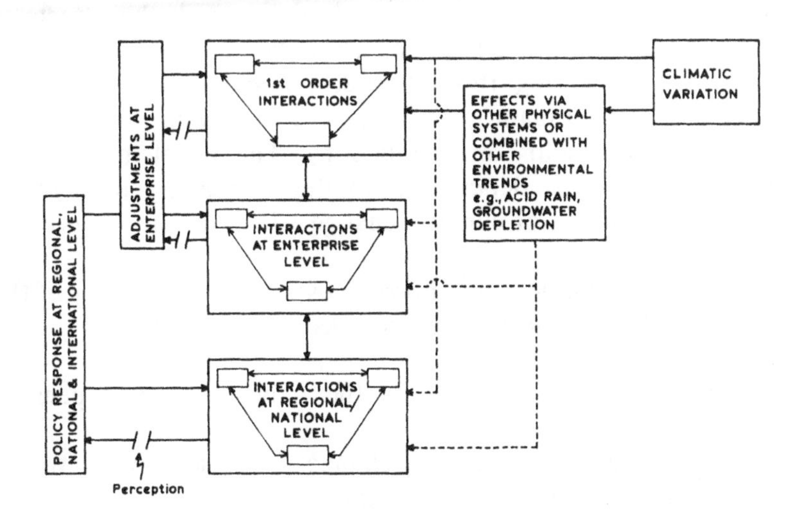

on employment, and on activity rates in non-agricultural sectors.

Secondly, the approach considers the effects of climatic changes and their interactions with other physical systems, distinguishing between:

- Those in which the effects are transmitted through other physical systems (e.g. by effects on pests and diseases; by changes in rates of soil erosion)
- Those in which the effects of climatic changes are themselves affected by concurrent environmental trends (such as acid deposition, groundwater depletion)

Thirdly, the approach considers two types of response to climate impacts:

- Technical adjustments at the farm level
- Policy responses at the regional, national and international level.

Adjoint versus direct methods

One disadvantage of the model hierarchy outlined above is that it suggests that a direct method is followed in which the effects of a change in climate are traced, in a number of steps, through a hierarchy of systems (e.g. through effects on plant growth, through the effects of altered plant growth rates on crop yield, through the effects of altered yields on production). The problem with this approach is that it is dependent on the availability of detailed and realistic scenarios of climatic change which are not at present available.

An alternative or *adjoint* method[4] focuses on the sensitivity of the exposure unit to climatic change and asks the questions:

- To what aspects of climate is the exposure unit especially sensitive?
- What magnitude and rate of change in these aspects would perturb the exposure unit significantly?

A schema of the direct and adjoint methods is given in Figure 3.5.

The advantage of the adjoint approach is that it can help distinguish between changes of climate that are significant and those that are trivial, and thus help identify target rates of climatic change to which policies of mitigation should be directed. To illustrate, if it were possible to identify magnitudes and rates of climatic change that could just be tolerated by a given type of agriculture in a certain region then this, together with a knowledge of other tolerable rates/magnitudes for other activities and ecosystems in other regions, can provide an indication of the scale of climatic change that might well be worth avoiding by policies of mitigation. The relative benefits of avoidance would, of course, depend on an assessment of the costs and benefits of mitigation versus adaptation.

Figure 3.5 a) Direct and b) adjoint methods of climate impact assessment. Both approaches were used in the IIASA/UNEP project. (*Source:* Parry & Carter, 1988).[4]

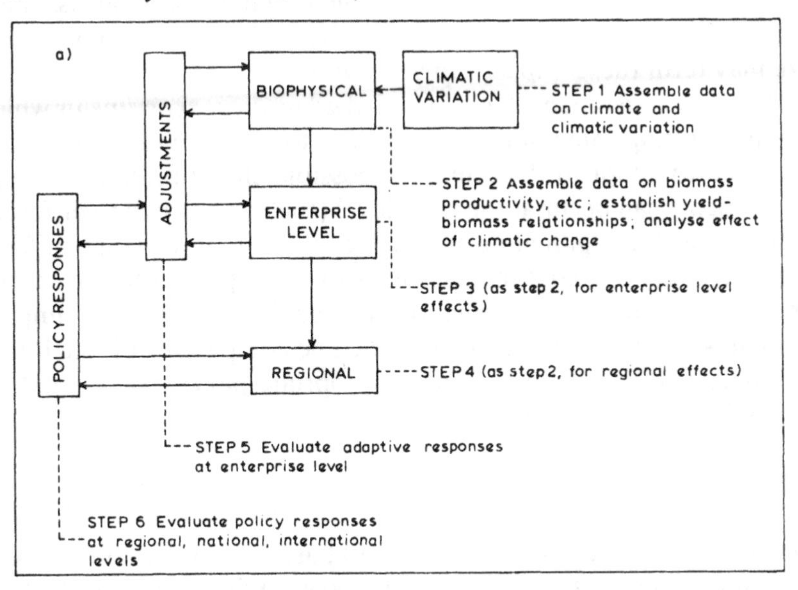

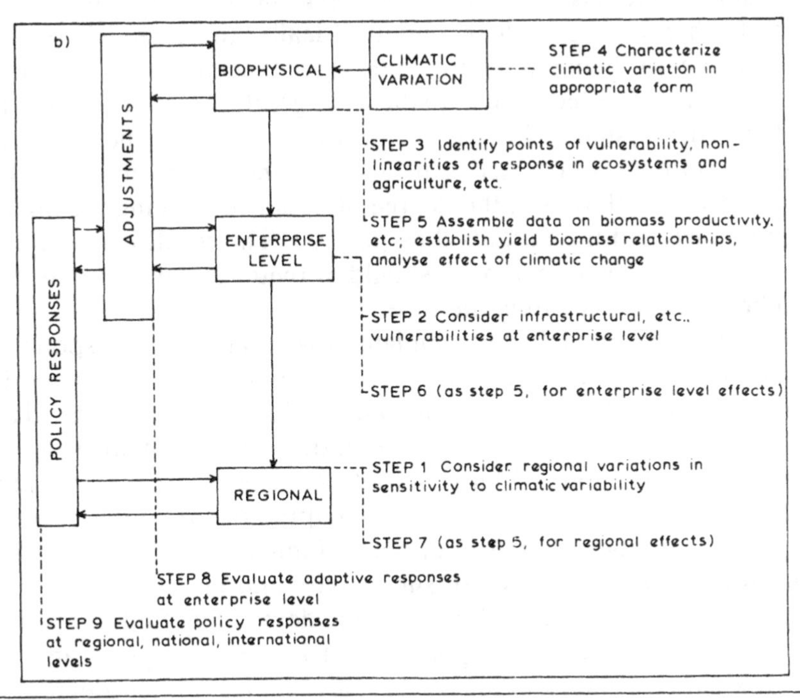

Impact experiments and adjustments

Until recently impact assessments have considered only a small number of feedback effects. Their analysis has generally been sequential, estimates of effects being based largely on assumed and essentially static sets of agronomic and economic responses. We may term these *impact experiments*.[4]

Very few assessments have included *adjustment experiments* which involve altering some of the assumptions to evaluate various options available to adapt to the effects of possible climatic change. These adjustments can occur at the farm level (such as a switch to different crops) or at the regional or national level (such as a change in farm support policy). Experiments with (for example) different crops, or different amounts of irrigation, can enable a new set of impact estimates to be generated that can then be compared with the initial estimates and thus help evaluate appropriate policies of adaptation to possible climatic changes.

THE CHOICE OF MODELS

The approaches outlined above are all based upon the effective use of a range of models – of climate, of crop growth, of food production and of other sectors in the economy. Ideally, these models should be capable of being linked together so that they can simulate the cascade of effects of climatic change through first-order to higher-order systems. In reality, however, few such models have been developed and tested satisfactorily and, more importantly, they are very different in the degree of detail concerning the inputs they require and the outputs they provide. The result is that while the outputs from one model should provide suitable inputs for another, this is very rarely the case.

For example, models that simulate how certain crops respond to variations of weather may provide information on variations in grain weight that is not the equivalent of marketable yield, data which are required as an input to models that can provide an estimate of variations of farm output. As a result, it is frequently necessary to interpret and alter the outputs of first-order models to provide suitable second-order model inputs. This process clearly introduces a further degree of subjectivity and inaccuracy into the analysis and can make highly detailed, demanding and expensive first-order modelling rather pointless. It is therefore preferable to adopt a suite of models that are broadly similar in their degree of resolution and are compatible with the degrees of uncertainty and inaccuracy inherent elsewhere in the analysis.

First-order (agroclimatic) models

There are two general techniques for examining the response of agricultural crops to climatic changes: the measurement of crop suitability by agroclimatic indices, and the estimation of potential productivity by crop-climate modelling.

a) Agroclimatic indices

These are used to characterize the growth environment of a crop on the basis of climatic variables manipulated either singly or in combination. For example, a commonly derived variable to characterize thermal agroclimate is Effective Temperature Sum (ETS) (sometimes also referred to as accumulated temperature and usually measured in units of growing degree-days). This usually represents the summation of temperature during the growing period above some base temperature assumed to be critical for crop growth.

A commonly used index for characterizing moisture aspects of agroclimate is an index of precipitation effectiveness (such as that developed by Thornthwaite)[10] and of drought (e.g that developed by Palmer).[11] An advantage of all these indices is that they do not demand large amounts of detailed data, and can therefore be employed for large-area geographical assessments of agroclimatic potential based on long-term mean climate data across a network of climatological stations.[12] This is one reason for their frequent use in recent impact assessments (see Chapters 4 and 5).

b) Crop-climate models

There is a wide range of models that has been developed for a number of different purposes.[13] For our purposes, two broad classes of models may be distinguished: empirical-statistical models and simulation models.

EMPIRICAL-STATISTICAL MODELS are based on the statistical relationship (assessed by techniques such as multiple regression analysis) between a sample of crop-yield data and a sample of weather data. This procedure has sometimes been labelled a "black-box" approach since it is not necessarily based on an understanding of the causal relationships between climate and crop yield (Figure 3.6). However, where empirical-statistical models are based on

Figure 3.6 Schema of the construction, operation and validation of a) an empirical-statistical and b) a simulation model. (*Source.* Carter *et al.* 1988).[12]

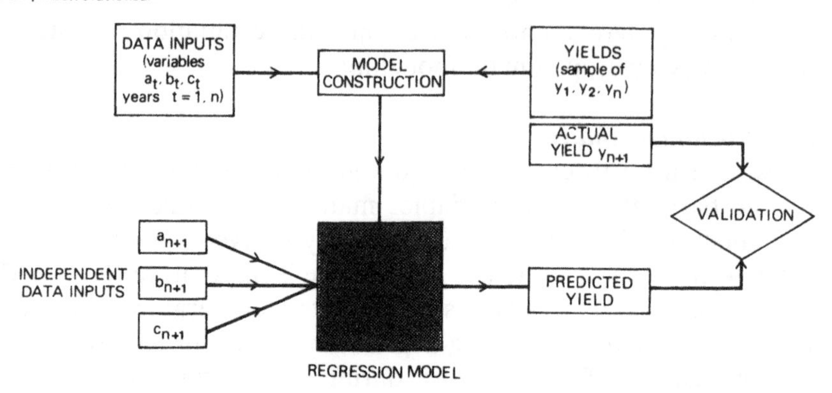

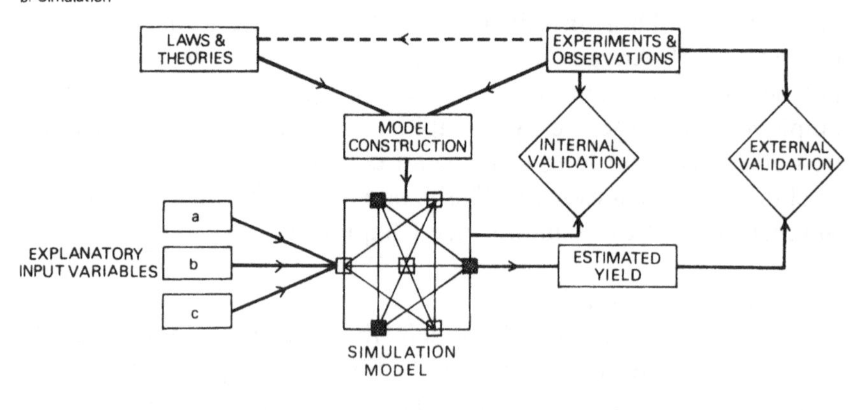

a good knowledge of crop physiology, and are thus the product of careful selection of suitable explanatory variables, they can be effective tools in climate-impact assessment.

SIMULATION MODELS express the dynamics of crop growth over the growing season through a set of mathematical equations that tie together the interrelationships of plant, soil and climate processes. When these are based on a close understanding of plant-growth processes they are the most accurate means of estimating plant responses to climatic change. One disadvantage, however, is that they frequently require detailed input data on climate, soils and management, and this has generally resulted in impact assessments based on them being restricted to a few data points covering a large area. A combination of models is probably the most effective means of analysis, with relatively simple agroclimatic indices being used to characterize the varying sensitivity of large areas to climatic change, thus identifying appropriate sample points for more detailed study with the use of crop–climate models. This approach has recently been adopted in a study of the impact of climatic change on agriculture in Europe funded by the European Commission.[14]

LINKING FIRST-ORDER AND HIGHER MODELS

A few studies have now been completed where output data from agroclimatic models have been used as input data to other (often economic) models that simulate the higher-order effects of climatic change. The latter include for example, models of output at the farm level, of land allocation to different crops, of employment, of food stocks, etc. A wide range of examples is given in Chapter 6.

DEVELOPING SCENARIOS OF CLIMATIC CHANGE

In order to estimate the likely effects of climatic changes on agriculture it is necessary to obtain a quantitative representation of the changes in climate themselves. These can then be used as inputs to the crop-climate models. Since climatic changes cannot yet be accurately predicted, the approach frequently adopted is to specify a set of plausible future conditions referred to as "scenarios". These climatic scenarios need to be meteorologically realistic, readily available or derivable and suitable as inputs to crop–climate models.

Most recent studies of climatic change have used $2 \times CO_2$ climatic scenarios based on outputs from GCMs (see Chapter 2). These are available for a network of grid points, varying between 4° and 8° latitude and 5°–10° longitude according to the particular GCM, showing the simulated change in daily or monthly averaged climatic variables (temperature, precipitation and cloud cover) between $1 \times CO_2$ (present-day or baseline climate) and $2 \times CO_2$ (future) equilibrium conditions.

Since, for the purposes of impact assessment, it is necessary to estimate effects at the regional level, it is important to evaluate the performance of the GCMs in simulating present-day regional climatic conditions. In fact the poor correspondence between real and modelled climate at the regional level for many parts of the Earth's surface, particularly for rainfall, means that little confidence can be attached to model estimates of regional climatic change. Since no alternative is available, however, most impact studies have had to assume that, while the GCMs do not reproduce the observed present-day climate very closely, the change between $1 \times CO_2$ and $2 \times CO_2$ equilibrium conditions indicates the probable difference between the present-day climate and a future $2 \times CO_2$ climate. The assumed changes in temperature and precipitation have generally been applied as differences to or ratios of the present-day climate.[15]

The time-dependent or transient response of climate to radiative forcing has, at present, only been considered in one set of GCM experiments (See Chapter 2). These results are now, however, providing the basis for a new generation of impact assessments.[14]

Some studies have sought also to use climatic scenarios based on historically analogous warm periods, using observed data from the meteorological record as inputs to crop-climate models.[15] Whether or not these are appropriate analogues is often most uncertain, but such studies also serve to estimate the current range of impacts on agriculture that stem from present-day variability of climate, and thus enable us to consider whether the expected effect over the longer term is within or exceeds the current range of experience.

Finally, some studies have adopted synthetic climatic scenarios that are generated specifically to simulate a climatic change. In conjunction with the adjoint method (see page 30 above), these have proved to be useful for estimating the amount of change in temperature and/or rainfall that is needed before a significant agricultural response occurs. In addition, in the absence of time-dependent transient GCM estimates, synthetic climatic scenarios may also be used to represent conditions intermediate between the present and $2 \times CO_2$ climates.

4. EFFECTS ON PLANTS, SOIL, PESTS AND DISEASES

There are three ways in which the Greenhouse Effect may be important for agriculture. First, increased atmospheric CO_2 concentrations can have a direct effect on the growth rate of crop plants and weeds. Secondly, CO_2-induced changes of climate may alter levels of temperature, rainfall and sunshine that can influence plant and animal productivity. Finally, rises in sea level may lead to loss of farmland by inundation and to increasing salinity of groundwater in coastal areas. These three types of impact will be considered in turn.

EFFECTS OF CO_2 ENRICHMENT

Effects on photosynthesis

If increases in atmospheric CO_2 were occurring without the possibility of associated changes in climate then, overall, the consequences for agriculture would probably be beneficial. CO_2 is vital for photosynthesis, and the evidence is that increases in CO_2 concentration would increase the rate of plant growth. Photosynthesis is the net accumulation of carbohydrates formed by the uptake of CO_2, so it increases with increasing CO_2. A doubling of CO_2 may increase the photosynthetic rate by 30 to 100%, depending on other environmental conditions such as temperature and available moisture.[1] More CO_2 enters the leaves of plants due to the increased gradient of CO_2 between the external atmosphere and the air space inside the leaves. This leads to an increase in the CO_2 available to the plant for conversion into carbohydrate.[2] The difference between photosynthetic gain and loss of carbohydrate by respiration is the resultant growth.

There are, however, important differences between the photosynthetic mechanisms of different crop plants and hence in their response to increasing CO_2. Plant species with the C3 photosynthetic pathway (the first product in their biochemical sequence of reactions has three carbon atoms) use up some of the solar energy they absorb in a process known as photorespiration, in which a significant fraction of the CO_2 initially fixed into carbohydrates is reoxidized back to CO_2.[3] C3 species tend to respond positively to increased CO_2 because it tends to suppress rates of photorespiration (Figure 4.1). This has major implications for food production in a high-CO_2 world because some of the current major food staples, such as wheat, rice and soya bean, are C3 plants.

Figure 4.1 Typical photosynthesis response of plants to CO_2. Net photosynthesis of wheat is about 70 mg of CO_2 $dm^{-2}h^{-1}$ compared with maize (about 55 mg of CO_2 $dm^{-2}h^{-1}$) for equivalent light intensity (0.4 cal cm^{-2} min^{-1}). Maize is saturated at a lower CO_2 concentration (c.450 ppmv) than wheat (c.850 ppmv). (Adapted from Akita and Moss, 1973).[4]

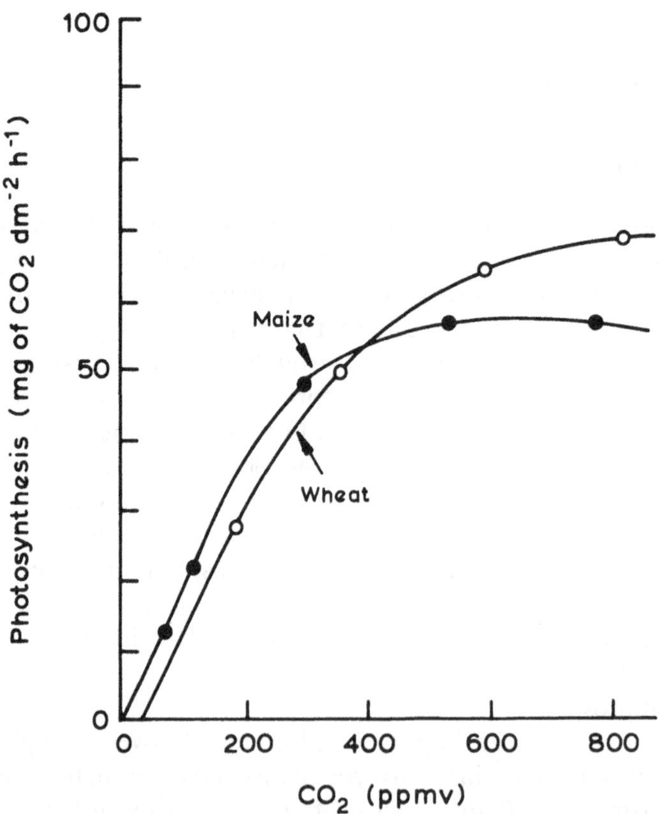

However, in C4 plants (those in which the first product has four carbon atoms) CO_2 is first trapped inside the leaf and then concentrated in the cells which perform the photosynthesis.[3] Although more efficient photosynthetically under current levels of CO_2, these plants are less responsive to increased CO_2 levels than C3 plants (Figure 4.1).

The major C4 staples are maize, sorghum, sugarcane and millet. Since these are largely tropical crops, and most widely grown in Africa, there is thus the suggestion that CO_2 enrichment will benefit temperate and humid tropical agriculture more than that in the semi-arid tropics and that, if the effects of climatic changes on agriculture in some parts of the semi-arid tropics are negative, then these may not be partially compensated by the beneficial effects of CO_2 enrichment as they might in other regions.

In addition we should note that, although C4 crops account for only about one-fifth of the the world's food production, maize alone accounts for 14 per cent of all production and about three-quarters of all traded grain. It is the major grain used to make up food deficits in famine-prone regions, and any reduction in its output could affect access to food in these areas.

C3 crops in temperate and subtropical regions could also benefit from reduced weed infestation. Fourteen of the world's 17 most troublesome terrestrial weed species are C4 plants in C3 crops.[5] The difference in response to increased CO_2 may make such weeds less competitive. In contrast, C3 weeds in C4 crops, particularly in tropical regions, could become more of a problem, although the final outcome will depend on the relative response of crops and weeds to climatic changes as well.

The different response of C3 and C4 crops may encourage changes in areas sown. It may, for example, accelerate the recent trend in India toward wheat, rice and barley and away from maize and millets, a trend that has largely been driven by the promise of greater increases in yield. It may tend to reverse the current trend in temperate areas away from perennial rye grass (a C3 crop) towards silage maize (C4) as the major forage crop; and in the USA it might encourage a tendency to switch from maize to soybean (C3) for forage.

Many of the pasture and forage grasses of the world are C4 plants, including important prairie grasses in North America and central Asia and in the tropics and subtropics.[6] The carrying capacity of the world's major rangelands are thus unlikely to benefit substantially from CO_2 enrichment. Much, of course, will depend on the parallel effects of climatic changes on the yield potential of these different crops.

The actual amount of increase in usable yield rather than of total plant matter that might occur as a result of increased photosynthetic rate is also problematic. In controlled environment studies, where temperature and moisture are optimal, the yield increase can be substantial, averaging 36 per cent for C3 cereals such as wheat, rice, barley and sunflower under a doubling of ambient CO_2 concentration (Table 4.1). Few studies have yet been made, however, of the effects of increasing CO_2 in combination with changes of temperature and rainfall.

Table 4.1: Mean predicted growth and yield increases for various groupings of C_3 species for a doubling of atmospheric CO_2 concentration from 330 ppmv to 660 ppmv. The errors indicated are 95% confidence limits.

	Footnote	*Immature crops*		*Mature crops*	
		No. of records	*% increase of biomass*	*No. of records*	*% increase of marketable yield*
Fibre crops	1	5	124	2	104
Fruit crops	2	15	40	12	21
Grain crops	3	6	20	15	36
Leaf crops	4	5	37	9	19
Pulses	5	18	43	13	17
Root crops	6	10	49	–	–
C3 weeds	7	10	34	–	–
Trees	8	14	26	–	–
Av. of all C_3		(83)	40±7	(51)	26±9

Source: Warrick *et al.*, 1986[7]

Footnotes: The species represented are:

1. cotton (*Gossypium hirsutum*);
2. cucumber (*Cucumis sativus*), eggplant (*Solanum melongena*), okra (*Abelmoschus esculentus*), pepper (*Capsicum annuum*), tomato (*Lycopersicum esculentum*);
3. barley (*Hordeum vulgare*), rice (*Oryza sativa*), sunflower (*Helianthus annuus*), wheat (*Triticum aestivum*);
4. cabbage (*Brassica oleracea*), *white clover* (*Trifolium repens*), fescue (*Festuca clatior*), lettuce (*Lactuca sativa*), Swiss chard (*Beta vulgaris*);
5. bean (*Phaseolus vulgaris*), pea (*Pisum sativum*), soybean (*Glycine max*);
6. sugar beet (*Beta vulgaris*), radish (*Raphanus lativus*);
7. *Crotalaria spectabilis, Desmodium paniculatum,* jimson weed (*Datura stramonium*), pigweed (*Amaranthus retroflexus*), ragweed (*Ambrosia artemisiifolia*), sicklepod (*Cassia obtusifolia*), velvet leaf (*Abutilon theophasti*);
8. cotton (*Gossypium deltoides*).

Little is also known about possible changes in yield quality under increased CO_2. The nitrogen content of plants is likely to decrease, while the carbon content increases, implying reduced protein levels and reduced nutritional levels for livestock and humans. This, however, may also reduce the nutritional value of plants for pests, so that they need to consume more to obtain their required protein intake.

Effects on water use by plants

Just as important may be the effect that increased CO_2 has on the closure of stomata, small openings in leaf surfaces through which CO_2 is absorbed and through which water vapour is released by transpiration. This tends to reduce the water requirements of plants by reducing transpiration (per unit leaf area) thus improving what is termed water use efficiency (the ratio of crop-biomass accumulation to the water used in evapotranspiration). A doubling of ambient CO_2 concentration causes about a 40 per cent decrease in stomatal aperture in both C3 and C4 plants[8] which may reduce transpiration by 23–46 per cent.[9] This might well help plants in environments where moisture currently limits growth, such as in semi-arid regions, but there remain many uncertainties, such as how much the greater leaf area of plants due to increased CO_2 will balance the reduced transpiration per unit leaf area.[10]

In summary, we can expect a doubling of atmospheric CO_2 concentrations from 330 to 660 ppmv to cause a 10 to 50 per cent increase in growth and yield of C3 crops (such as wheat, soybean and rice) and a 0 to 10 per cent increase for C4 crops (such as maize and sugarcane).[7] Much depends, however, on the prevailing growing conditions. Our present knowledge is based on a few experiments mainly in glass-houses and has not yet included extensive study of response in the field under subtropical conditions. Thus, although there are indications that, overall, the effects of increased CO_2 could be distinctly beneficial and could partly compensate for some of the negative effects of CO_2-induced changes of climate, we cannot at present be sure that this will be so.

EFFECTS OF INCREASED TEMPERATURES

Effects on growth-rates

In high mid-latitude regions (above 45°), at high latitudes (above 60°) and at high altitudes, temperature is frequently the dominant climatic control on crop and animal growth. It determines the potential length of the growing and grazing seasons, and generally has a strong effect on the timing of developmental processes and on rates of expansion of plant leaves. The latter, in turn, affects the time at which a crop canopy can begin to intercept solar radiation and thus the efficiency with which solar radiation is used to make plant biomass.[11]

In general, plant response to temperature follows that indicated in Figure 4.2. Development does not begin until temperature exceeds a threshold; then the rate of development increases broadly linearly with temperature to an optimum, above which it decreases broadly linearly.[12]

However, the effect of this development on plant biomass depends on whether the growth habit of the plant is determinate (that is, it has a discrete life cycle which ends when the grain is mature, such as in cereals), or whether it is indeterminate (that is, it continues to grow and yield throughout the season, such as in grasses and

Figure 4.2 Temperature, development and canopy expansion. (a) Idealized relation between developmental rate and temperature. Development does not begin until temperature exceeds a threshold (T_b, the base temperature); then developmental rate increases linearly with temperature to an optimum (T_o), above which it decreases linearly. (b) and (c) Effect of temperature on the relation between time and fractional interception of solar radiation by a canopy, for a determinate (b), and indeterminate (c) species. (----- cooler; —, warmer). (*Source:* Squire and Unsworth, 1988).[12]

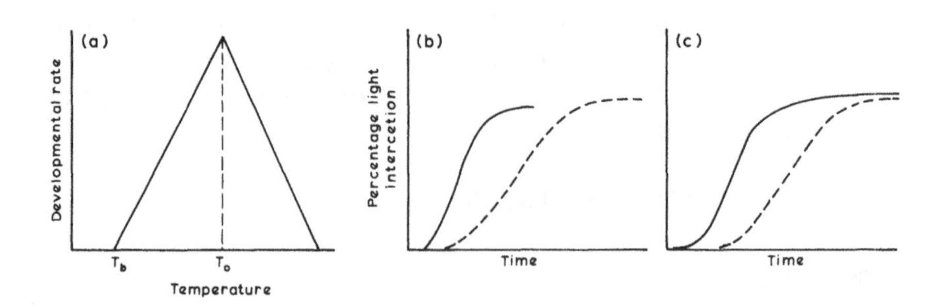

rootcrops). Temperature increase shortens the reproductive phase of determinate crops, decreasing the time during which the canopy exists and thus the period during which it intercepts light and produces biomass (Figure 4.2b). The canopy of indeterminate crops, however, continues to intercept light until it is reduced by other events such as frost or pests, and the duration of the canopy increases when increased temperatures extend the season over which crops can grow (e.g., by delaying the first frosts of autumn) (Figure 4.2c). An increase in temperature above the base but not exceeding optimum temperatures should therefore generally lead to lower yields in cereals and higher yields of root crops and grassland, though higher temperatures may also lead to higher rates of evaporation and therefore reduced moisture availability that can also be expected to affect yields. These effects on moisture are discussed later.

Effects on growing seasons

One of the most important effects of an increase in temperature, particularly in regions where agricultural production is currently limited by temperature, would be to extend the growing season available for plants (e.g. between last frost in spring and first frost in autumn) and reduce the growing period required by crops for maturation. An example is given, for the Canadian prairies, in Figure 4.3. Here the length of growing season is estimated to increase by about 10 days per °C increase in mean annual temperature. At the same time the maturation time for spring wheat is reduced by about 3 days per °C, with the result that the probability of the crop not maturing before first autumn frost is reduced by as much as 5 per cent per °C.[13] However, the length of time during which the crop is producing dry matter (from heading to ripening in Figure 4.2c) is also reduced with consequently reduced average grain yield. In the Canadian prairies warming therefore implies less frost damage but lower average yields of spring wheat.

The effects of warming on length of growing season and growing period will vary from region to region and from crop to crop. For wheat in Europe, for example, the growing season is estimated to lengthen by about 10 days per °C and in central Japan by about 8 days per °C.[14, 15] In general the conclusion is that increased mean annual temperatures, if limited to two or three degrees, could generally be expected to extend growing seasons in high mid-latitude and high-latitude regions. Increases of more than this could increase evapotranspiration rates to a point where reduced crop-water

Figure 4.3 Effects of temperature deviations from 1951–80 normal on: (a) changes in growing season length for crop districts 1a, 9a and a provincial average for Saskatchewan, (b) changes in maturation time for spring wheat, and (c) changes in spring wheat phase development lengths. (*Source:* Williams *et al.* 1988).[13]

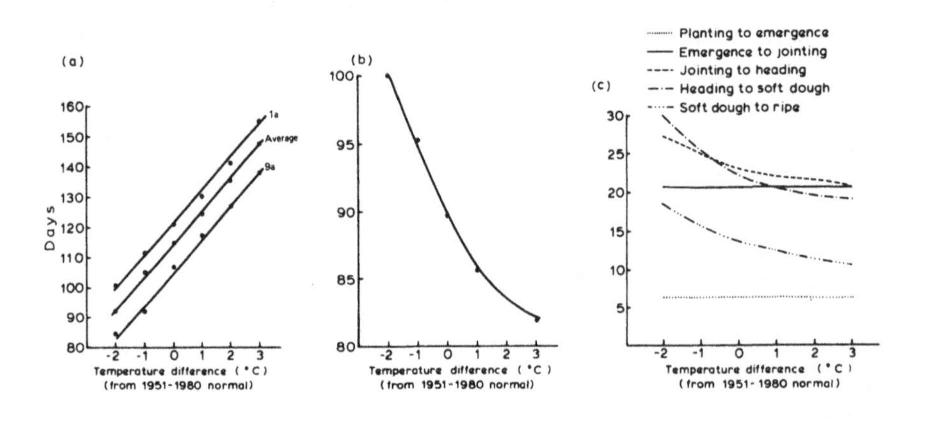

availability begins to limit the growing season. The effects of these changes in growing season on agricultural potential are discussed in the following chapter.

Effects on yield

Whether crops respond to higher temperatures with an increase or decrease in yield depends on whether they are determinate or indeterminate, and whether their yield is currently strongly limited by insufficient warmth. In cold regions very near the present-day limit to arable agriculture any temperature increase, even as much as the 7–9°C indicated for high latitudes under a doubling of CO_2 (see Chapter 2), can be expected to enhance yields of cereal crops. For example, near the current northern limit of spring-wheat production in the European region of the USSR yields increase about 3 per cent per °C, assuming no concurrent change in rainfall (Table 4.2). In Finland, the marketable yield of barley increases 3–5 per cent per °C, and in Iceland hay yields increase about 15 per cent per °C.[17, 18]

Away from current temperature-constrained regions of farming

Table 4.2: Response of spring-wheat yield (as percentages of the long-term mean) to variations in air temperature (ΔT) and rainfall (ΔR) during the growing season (Cherdyn, forest zone).

	ΔT (°C)		
ΔR (mm)	*−1.0*	*0*	*+1.0*
−20	93	97	99
0	95	100	103
+20	97	101	107

Source: Pitovranov *et al.*, 1988[16]

and in the core areas of present-day cereal production such as in the Corn Belt of North America, the European lowlands and the Soviet Ukraine, increases in temperature would probably lead to decreased cereal yield due to a shortened period of crop development.[19] In eastern England, for example, a 3°C rise in mean annual temperature is estimated to reduce winter-wheat yield by about 10 per cent although the direct effect of a doubling of ambient atmospheric CO_2 might more than compensate for this (Figure 4.4).

In other mid-latitude regions much would depend on possible changes in rainfall. For example, in the Volgograd region, just east of the Ukraine, spring wheat yields are estimated to fall only a small amount with a 1°C increase in mean temperature during the growing season, though they could increase or decrease substantially if the temperature change was accompanied by an increase or decrease of rainfall (Table 4.3).

Table 4.3: Response of spring wheat yield (as percentages of the long-term mean) to variations in air temperature (ΔT) and precipitation (ΔP) during the growing season (Palassovka, Volgograd region).

	ΔT (°C)				
ΔP (mm)	−1.0	−0.5	0	+0.5	+1.0
−40	79	79	76	76	76
−20	92	92	89	89	89
0	104	103	100	100	99
+20	115	114	110	109	108
+40	125	124	120	118	117

Source: Nikonov *et al.*, 1988[20]

Yields of root crops such as sugar beet and potatoes, with an indeterminate growth habit, can be expected to see an increase in yield with increasing temperatures, provided these do not exceed temperatures optimal for crop development.[12]

Effects on livestock

A rise in temperature could also have a significant effect on the performance of farm animals, in addition to the effects that might

Figure 4.4 Modelled responses of total dry matter production and grain yield of winter wheat. A = curves modelled from the 1981 climatic conditions at Brooms Barn, Bury St Edmunds (UK); B, simulates the effect of a doubling of carbon dioxide concentration; and C, the effect of both a doubled carbon dioxide concentration and a rise in mean temperature of 3°C. (*Source:* Squire and Unsworth, 1988).[12]

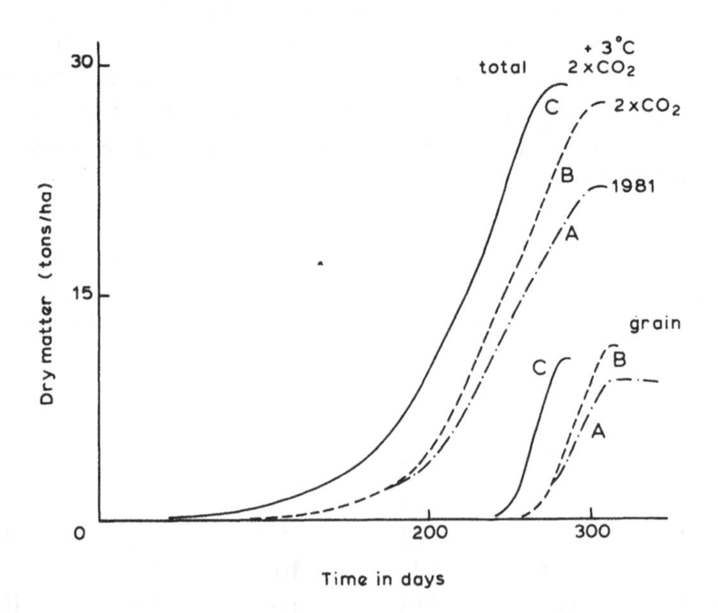

flow from altered yields of grassland and forage crops. Young animals tend to be less tolerant of a wide range of temperature than adults (Figure 4.5). A rise in summer temperatures, especially in regions with a continental climate characterized today by summer temperatures near the threshold tolerated by livestock (such as the south-central USA and USSR) could be detrimental to production.[12]

Figure 4.5 Temperature zones in which farm animals perform effectively. Numbers alongside boxes indicate temperature range. (adapted from: Bianca, 1976).[21]

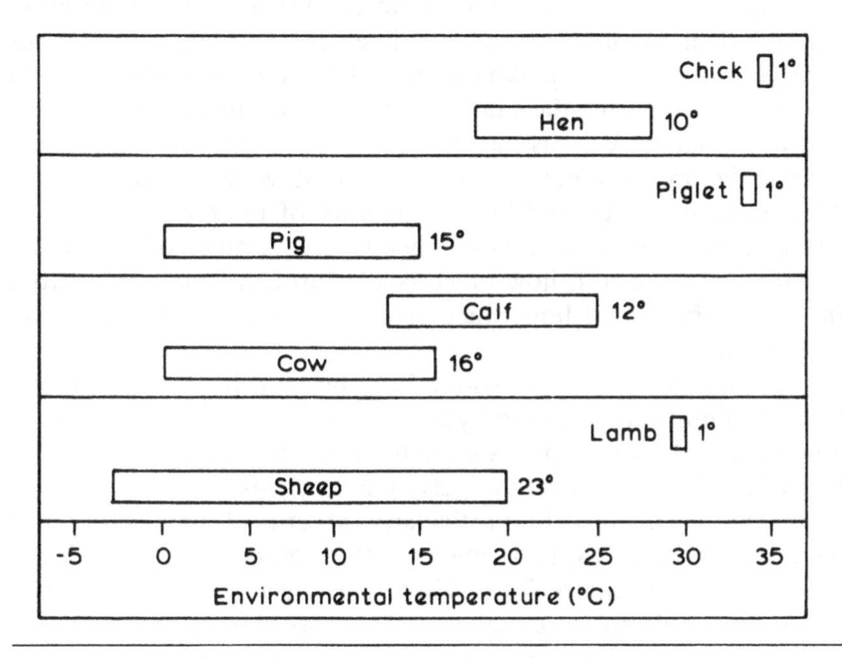

Effects on moisture availability

Changes of temperature would also have an effect on moisture available for crop growth, whether or not levels of rainfall remained unchanged. In general, and at mid-latitudes, evaporation increases by about 5 per cent for each °C of mean annual temperature. Thus, if mean temperature were to increase in the east of England by 2°C potential evaporation would increase by about 9 per cent (assuming no change in rainfall). The effect of this would be small in the early part of the growing season, but after mid-July the soil moisture deficit would be considerably larger than at present and, for some crops, this implies substantially increased demand for irrigation.[22] Of course, the amount of water available for plant growth is affected by a combination of climatic and non-climatic variables such as precipitation, temperature, sunshine, windspeed as well as soil porosity, slope, etc. These are considered in the following section.

EFFECTS OF CHANGES IN SOIL MOISTURE

In most of the tropical and equatorial regions of the world, and across large areas outside the tropics, the yield of agricultural crops is limited more by the amount of water received by and stored in the soil than by air temperature. Even in the high mid-latitudes such as in southern Scandinavia too little rain can restrict growth of cereal crops during the summer when evapotranspiration exceeds rainfall. In all these areas the amount of dry matter a crop produces is roughly proportional to the amount of water it transpires.[11] This, in turn, is affected by the quantity of rainfall but not in a straightforward manner: it also depends on how much of the rainfall is retained in the soil, how much is lost through evaporation from the soil surface, and how much remains in the soil that the crop cannot extract.

The amount of water transpired by the crop is also determined by air humidity, with generally less dry matter produced in a drier atmosphere.[11] Thus, changes in both rainfall and air humidity would be likely to have significant effects on crop yields.

Reliability of rainfall, particularly at critical phases of crop development, can explain much of the variation in agricultural potential in tropical regions. Thus, many schemes used to map zones of agricultural potential around the world have adopted some form of ratio of rainfall to potential evaporation, r/Eo, to delimit moisture-availability zones, which are then overlaid on temperature and soils maps to indicate agro-ecological zones.[23] The regions are distinguished largely on the basis of the length of growing season determined by the r/Eo ratio. In Kenya, for example, average plant biomass is estimated to vary by more than an order of magnitude between agroclimatic zones that lie within 100 km of each other.[24] These are characterizations of the effect of differences in *average* rainfall on agricultural potential, but it is important to note that a high degree of inter-annual variability of rainfall, particularly in the drier zones, can lead to very marked variation in crop yield between wet and dry years, so that changes in rainfall over time as well as over space are also likely to have a similar effect on crop yields.

A strongly positive relationship between rainfall and crop yield is generally found in the major mid-latitude cereal-exporting regions of the world, such as the US Great Plains and Soviet Ukraine. For example, in the dry steppe zone of the Volga Basin (USSR), a 0.5 or 1°C warming, with no change in rainfall, is estimated to have little effect on spring-wheat yields, while a 20 per cent decrease in rainfall (at current temperatures) could reduce yields by more than a tenth (Table 4.3).

Relatively few studies have been made of the combined effects of possible changes in temperature and rainfall on crop yields, and those that have are based on a variety of different methods. However, a recent review of results from about ten studies in North America and Europe noted that warming is generally detrimental to yields of wheat and maize in these mid-latitude core cropping regions. With no change in precipitation (or radiation) slight warming (+1°C) might decrease average yields by about 5 ± 4 per cent; and a 2°C warming might reduce average yields by about 10 ± 7 per cent.[7] In addition, reduced precipitation might also decrease yields of wheat and maize in these breadbasket regions. A combination of increased temperatures (+2°C) and reduced precipitation could lower average yields by over a fifth.

EFFECTS ON IMPACTS FROM CLIMATIC EXTREMES

Important effects from changes of climate need not only stem from changes in average temperature and rainfall, but also from changes in the frequency of extreme climatic events that can be damaging and costly for agriculture. The balance between profit and loss in commercial farming often depends on the relative frequencies of favourable and adverse weather; for example, on the Canadian prairies a major constraint on profitable wheat production is related to the probability of the first autumn frost occurring before the crop matures.[25]

Among semi-commercial and subsistence farmers the probability of yield in a given year being more than a minimum necessary to feed the household may be more important than the average over several years.[26] Levels of risk such as these may well be altered quite markedly by apparently small changes in mean climate, particularly the risk of successive extremes, which can quickly lead to famine in food-deficit regions.

To illustrate, suppose that extremely dry summers (of a kind that can cause severe food shortage in a given region) occur at present with a probability of $P = 0.1$. The return period of the occurrence of a single drought is, therefore, 10 years, while the return period for the occurrence of two successive droughts is 100 years (assuming a normal distribution of frequencies). A change in climate can lead to a change in *P*, either through altered variability which will change *P* directly, and/or through a change in mean conditions that must also change *P* if drought is judged relative to an absolute threshold. Alternatively, *P* may change through changes in some critical impact threshold as a result of altered land use, increasing population pressure, and so forth. If *P* becomes 0.2, then the return period of a single drought is halved to 5 years. The return period for two successive droughts, however, is reduced by a factor of four to only 25 years.[26, 27] Thus, not only is agriculture often sensitive to climatic extremes, but the risk of climatic extremes may be very sensitive to relatively small changes in the mean climate.

The sensitivity of marginal farmers to climatic change may be especially great. The reason for this is that, near the margins of cultivation, the probability of critical levels of warmth or moisture required to avoid crop failure or a critical crop shortfall tends to increase not linearly but quasi-exponentially towards the margin of cultivation (Figure 4.6). Marginal areas are thus commonly characterized by a very steep "risk surface", with the result that any changes in average warmth or aridity, or in their variability, would have a marked effect on the level of risk in agriculture.

For the reasons given above, much of the impact on agriculture from climatic change can be expected to stem from the effects of extreme events. Consider, first, the significantly increased costs resulting from increased frequency of extremely hot days causing heat stress in crops. In the central USA the number of days with temperatures above 35°C, particularly at the time of grain filling, has a significant negative effect on maize and wheat yields.[29, 30, 31] The incidence of these very hot days is likely to increase substantially with a quite small increase in mean temperature. For example, in Iowa, in the US Corn Belt, an increase in mean temperature of only 1.7°C may bring about a three-fold increase in the probability of 5 consecutive days with a maximum temperature over 35°C.[32] At the southern edge of the Corn Belt, where maize is already grown near its maximal temperature-tolerance limit, such an increase could have a very deleterious effect on yield.

Figure 4.6 Probability of crop "failure", net loss or critical shortfall, with linear (and normally distributed) gradient of aridity or warmth. Sample area is south Scotland, probabilities of crop failure are for oats (var. Blainslie). Probabilities of crop failure which define the marginal areas are derived from empirical data on farming strategies in south Scotland. (*Source:* Parry, 1976).[28]

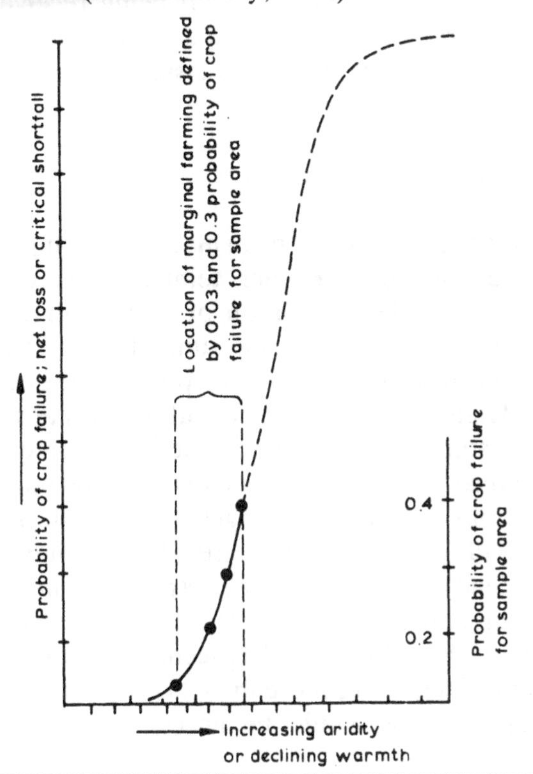

The increase in risk of heat stress on crops and livestock due to global warming could be especially important in tropical and subtropical regions where temperate cereals are currently grown near their limit of heat tolerance. For example, in northern India, where GCM experiments indicate an increase in mean annual temperature of about 4°C, wheat production might no longer be viable.

An important additional effect of warming, especially in temperate regions, is likely to be the reduction of winter chilling (vernalization). Many temperate crops require a period of low temperatures in winter either to initiate or to accelerate the flowering process. Low vernalization results in low flower-bud initiation and, ultimately, reduced yields. A 1°C warming could reduce effective winter chilling by between 10 and 30 per cent.[33]

Changes in rainfall could have a similarly magnified impact. For example, if mean rainfall in the Corn Belt in March (which is about 100 mm [4 inches]) decreased by 10 per cent (an amount projected by some GCMs under a 2 × CO_2 climate) this would raise the probability of less than 25 mm [1 inch] being received by 46 per cent. For cattle, crops and trees a 1 per cent reduction in rainfall could mean that drought-related yield losses increase by as much as a half.[34]

EFFECTS ON SOIL FERTILITY AND EROSION

No comprehensive study has yet been made of the impact of possible climatic changes on soils. Higher temperatures could increase the rate of microbial decomposition of organic matter, adversely affecting soil fertility in the long run.[3] But increases in root biomass resulting from higher rates of photosynthesis could offset these effects. Higher temperatures could accelerate the cycling of nutrients in the soil, and more rapid root formation could promote more nitrogen fixation. But these benefits could be minor compared to the deleterious effects of changes in rainfall. For example, increased rainfall in regions that are already moist could lead to increased leaching of minerals, especially nitrates. In the Leningrad region of the USSR a one-third increase in rainfall (which is consistent with the GISS 2 × CO_2 scenario) is estimated to lead to falls in soil productivity of more than 20 per cent. Large increases in fertilizer applications would be necessary to restore productivity levels.[16]

Decreases in rainfall, particularly during summer, could have a more dramatic effect, through the increased frequency of dry spells leading to increased proneness to wind erosion. Susceptibility to wind erosion depends in part on cohesiveness of the soil (which is affected by precipitation effectiveness) and wind velocity. The only study completed on this subject suggests that in Saskatchewan (on the Canadian prairies) the frequency of moderate and extreme droughts would increase three-fold under a 2 × CO_2 climate if mean May–August temperatures increased by 3.5°C and precipitation increased by 9 to 14 per cent, which is consistent with the GISS 2 × CO_2 climate. They would increase 13-fold if increases in temperature are not accompanied by increases in precipitation.

Estimated changes in the potential for wind erosion under the latter scenario vary from +24 to +29%.[13]

EFFECTS ON PESTS AND DISEASES

Studies suggest that temperature increases may extend the geographic range of some insect pests currently limited by temperature.[35] Figure 4.7 shows the results from one of the first of these studies – an assessment of the effects of climatic change on the potential distribution of the European Corn Borer (*Ostrinia nubilalis*) in Europe.[36] The European Corn Borer is a major pest of grain maize in many parts of the world. It is multivoltine (multigenerational) and, depending on climatic conditions, can produce up to four generations per year. Using degree-day (thermal) requirements, the potential distribution of the European Corn Borer in Europe has been mapped under present (1951–80) temperatures. With a 1°C increase in temperature a northward shift in distribution of between 165 and 500 km is indicated for all generations. In addition to favourable climatic conditions the distribution of any pest is dependent on the availability of a host plant. As indicated in Figure 5.3 the potential limit of grain maize cultivation is also likely to shift northwards with increasing temperatures providing suitable conditions for the European Corn Borer. This example serves to highlight the need to examine crop-pest interactions in any climate impact assessment.

Under a warmer climate at mid-latitudes there would be an increase in the overwintering range and population density of a number of important agricultural pests, such as the potato leafhopper which is a serious pest of soybeans and other crops in the USA.[19] Assuming planting dates did not change, warmer temperatures would lead to invasions earlier in the growing season (i.e. at more susceptible stages of plant development) and probably lead to greater damage to crops. In the US Corn Belt increased damage to soybeans is also expected due to earlier infestation by the corn earworm, which could result in serious economic losses.

Examination of the effect of climatic warming on the distribution of livestock diseases suggests that those at present limited to tropical countries, such as Rift Valley fever and African Swine fever, may spread into the USA causing serious economic losses.[19] The geographic distribution and activities of other diseases already important in the USA may also expand. The horn fly, which currently causes losses of $730.3 million in the beef and dairy

Figure 4.7 Potential number of generations of the European Corn Borer (*ostrinia nubilalis*) for a) present-day climate (1951–80) and b) a 1°C increase in mean annual temperature. (*Source*: J.H. Porter, personal communication, 1990).[36]

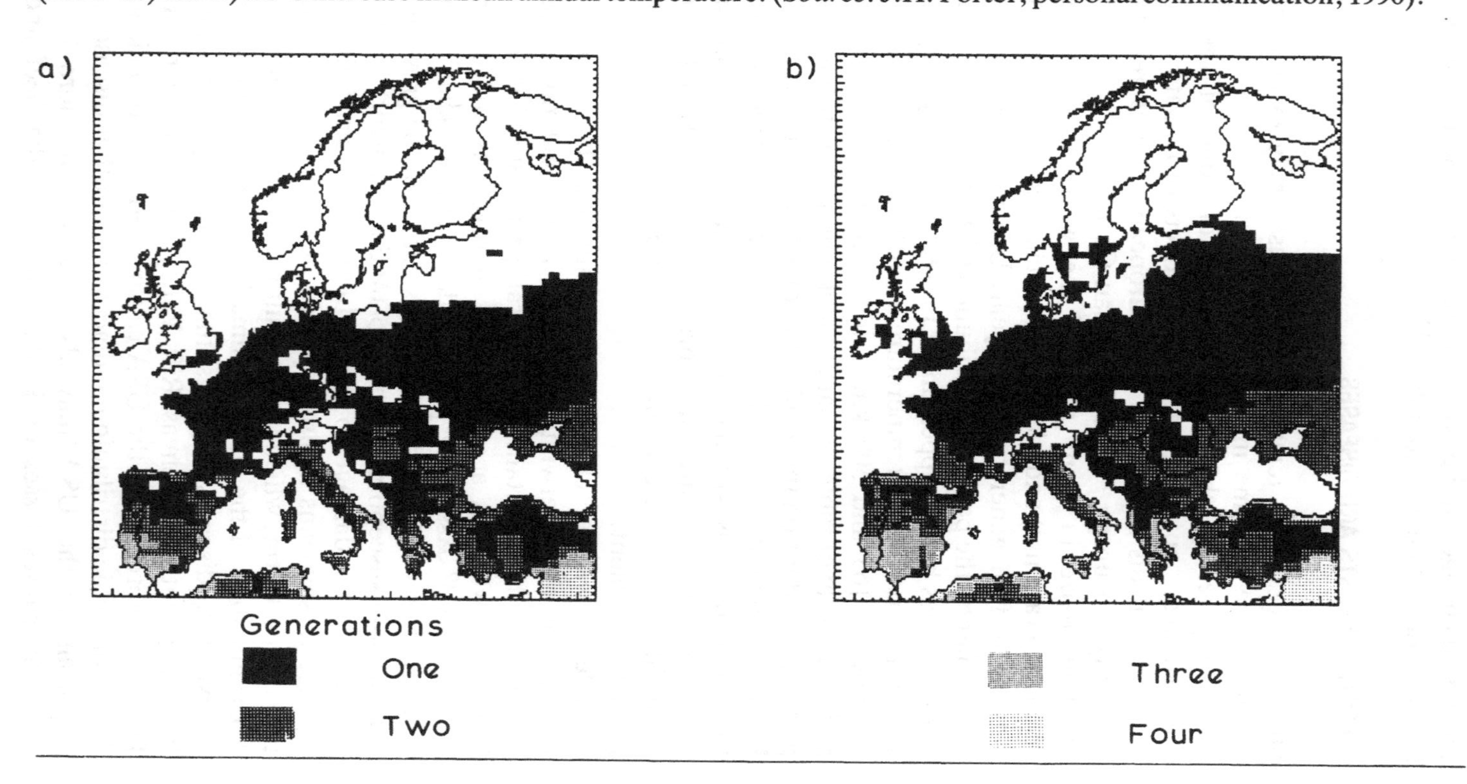

cattle industries might extend its range under a warmer climate leading to reduced gain in beef cattle and a significant reduction in milk production.[19, 37] In the 1960s and 1970s a combination of the increased resistance of ticks to insecticides and the high costs of dipping threatened the profitability of the Australian beef industry. Prolonged summer rainfall and an extended developmental season, or, conversely, prolonged dryness leading to increased nutritional stress in the host, are likely to cause heavy infestations.[38] If such climatic conditions were to prevail in the future it is likely that ticks could become an increasing problem.

One of the major threats of climatic change is the establishment of "new" or migrant pests as climatic conditions become more favourable to them. In New Zealand, for example, the swarming of locusts in the North Island in recent years may be an indication of a more widespread problem in the future.[39] In a similar fashion, anomalously warm conditions in 1986–1988 led to locust swarms reaching new limits in southern Europe.[40]

In cool temperate regions, where insect pests and diseases are not generally serious at present, damage is likely to increase under warmer conditions. In Iceland, for example, potato blight currently does little damage to potato crops, being limited by the low summer temperatures. However, under a $2 \times CO_2$ climate that may be 4°C warmer than at present, crop losses to disease may increase to 15 per cent.[18]

Most agricultural diseases have greater potential to reach severe levels under warmer conditions. Fungal and bacterial pathogens are also likely to increase in severity in areas where precipitation increases.[41] Under warmer and more humid conditions cereals would be more prone to diseases such as Septoria. In addition, increases in population levels of disease vectors could lead to increased epidemics of the diseases they carry. To illustrate, increases in infestations of the Bird Cherry aphid (*Rhopalosiphum padi*) or Grain aphid (*Sitobian avenae*) could lead to increased incidence of Barley Yellow Dwarf virus in cereals.

EFFECTS ON OTHER ECOSYSTEMS

It is possible that some of the impact of climatic changes on agriculture would stem not directly from the effects of altered temperature, precipitation, radiation, etc. on crops and animals, nor even indirectly from effects on pests, diseases and soils, but through potential effects on natural and semi-natural plant communities.

For example, if warming were to induce a northward shift of the boreal forest in northern regions of America, Europe and Asia, it is possible that extensive grazing, livestock rearing and cultivation of quick-maturing crops (farming types currently located at the southern limit of the boreal zone) would be encouraged to shift northwards to exploit regions vacated by forestry. A geographic shift of agriculture in these marginal regions would thus be the combined result of changes in potential for farming and changes in potential for forestry, with the outcome perhaps determined by the comparative advantage of one use over the other; and this might further be influenced by future policies of conservation.

An illustration of the possible extent of poleward shift of the boreal zone in the northern hemisphere is given in Figure 4.8. This is based on an estimation of the levels of effective temperature

Figure 4.8 Calculated boreal zone for the GISS 2 × CO_2 climate scenario relative to the calculated present-day zone. (*Source:* Kauppi & Posch, 1988).[42]

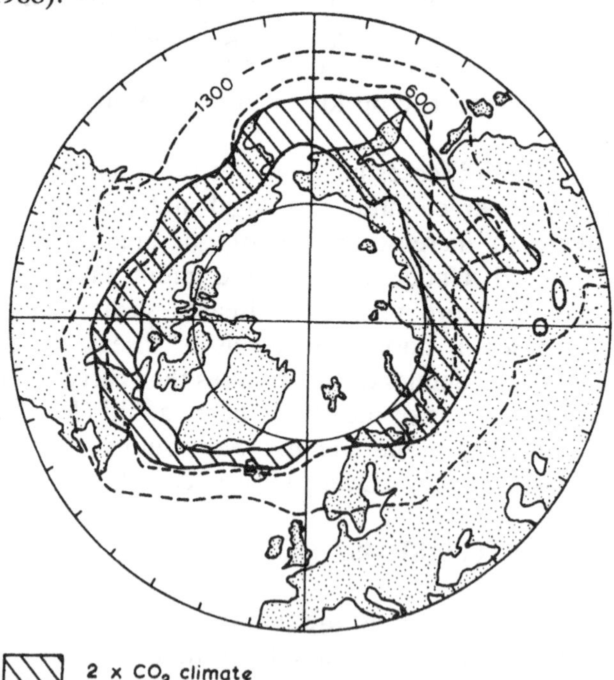

sum above a threshold temperature of 5°C that currently define the northern and southern limits of the boreal zone (600 and 1,300 degree-days, respectively).[42] Under the warming projected for a 2 × CO_2 climate (in this instance based on experiments with the GISS GCM) these limits are re-located about 500–1,000 km further north than at present. If taken as a proxy limit of northern agriculture, this indicates a substantial extension of agricultural potential, although much of this may be severely limited by inappropriate soils and terrain, particularly in northern America and Europe.

Similarly substantial shifts can be expected to occur for vegetation zones throughout the world. An illustration of the possible scale of these shifts, in this instance estimated for Europe for a 5°C increase in mean annual temperature and a 10 per cent increase in precipitation is given in Figure 4.9. On average, the major vegetation zones shift northwards by 1,000 km, the largest changes being in the boreal and Mediterranean regions.[43]

EFFECTS OF SEA-LEVEL RISE ON AGRICULTURE

CO_2-induced warming is expected to lead to rises in sea level as a result of thermal expansion of the oceans and partial melting of glaciers and ice caps, and this in turn is expected to affect agriculture, mainly through the inundation of low-lying farmland but also through the increased salinity of coastal groundwater. The IPCC estimate of sea-level rise above present levels under the Business-As-Usual scenario is 9 cm – 29 cm by the year 2030 with a best estimate of 18 cm, and 28 cm – 96 cm by 2090, with a best estimate of 58 cm.[1, 44]

Preliminary surveys of proneness to inundation have been based on a study of existing contoured topographic maps, in conjunction with knowledge of the local "wave climate" that varies between different coastlines. They have indentified 27 countries as being especially vulnerable to sea-level rise, on the basis of the extent of land liable to inundation, the population at risk and the capability of taking protective measures.[45] It should be emphasized, however, that these surveys assume a much larger rise in sea levels (1.5 m) than is at present estimated to occur within the next century under current trends of increase of GHG concentrations. On an ascending scale of vulnerability (1 to 10) experts identified the following most vulnerable countries or regions: 10, Bangladesh; 9, Egypt, Thailand; 8, China; 7, western Denmark; 6, Louisiana; 4, Indonesia.

The most severe effects on agriculture are likely to stem directly

Figure 4.9 Potential natural vegetation map for Europe based a) on current average temperature and precipitation and b) on average

a)

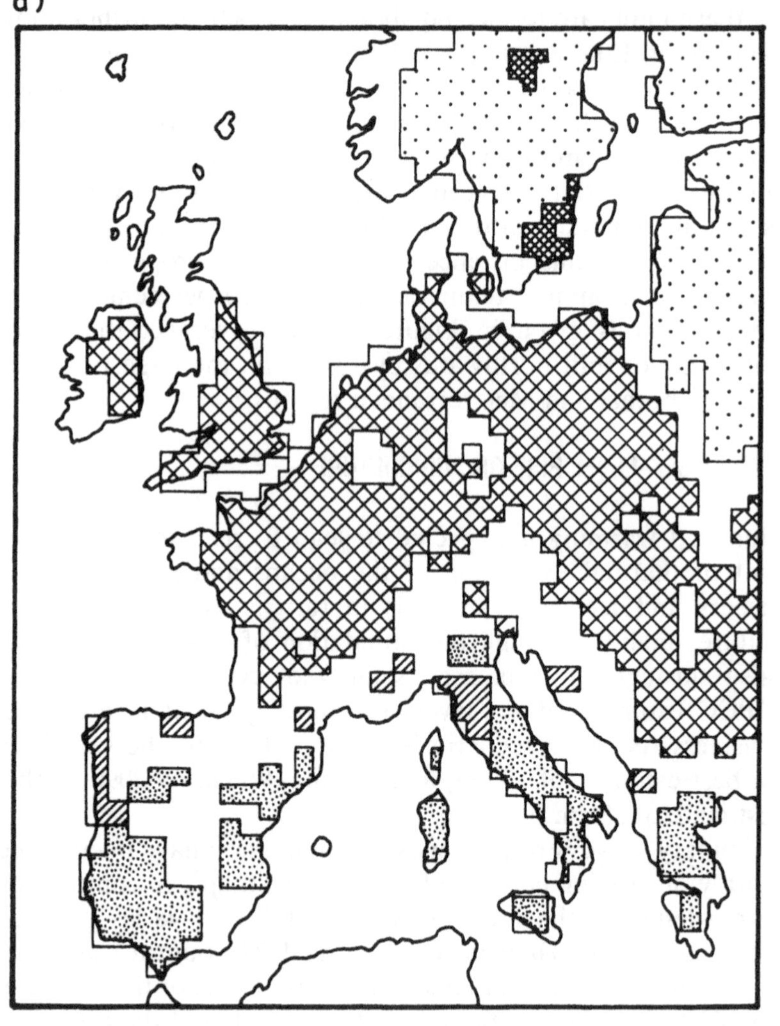

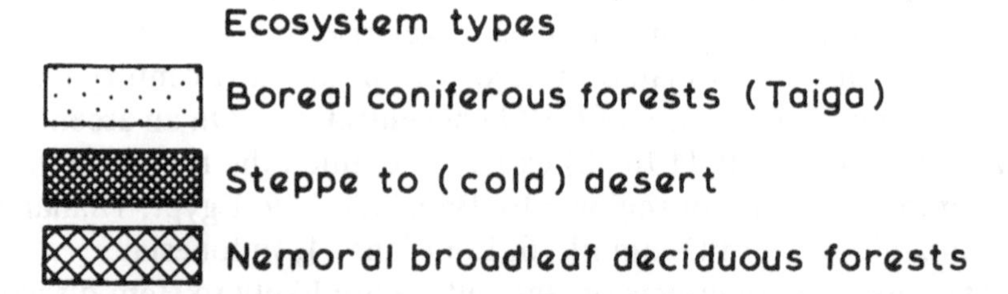

temperatures of +5°C and average precipitation of +10%. (*Source:* De Groot, 1987).[43]

b)

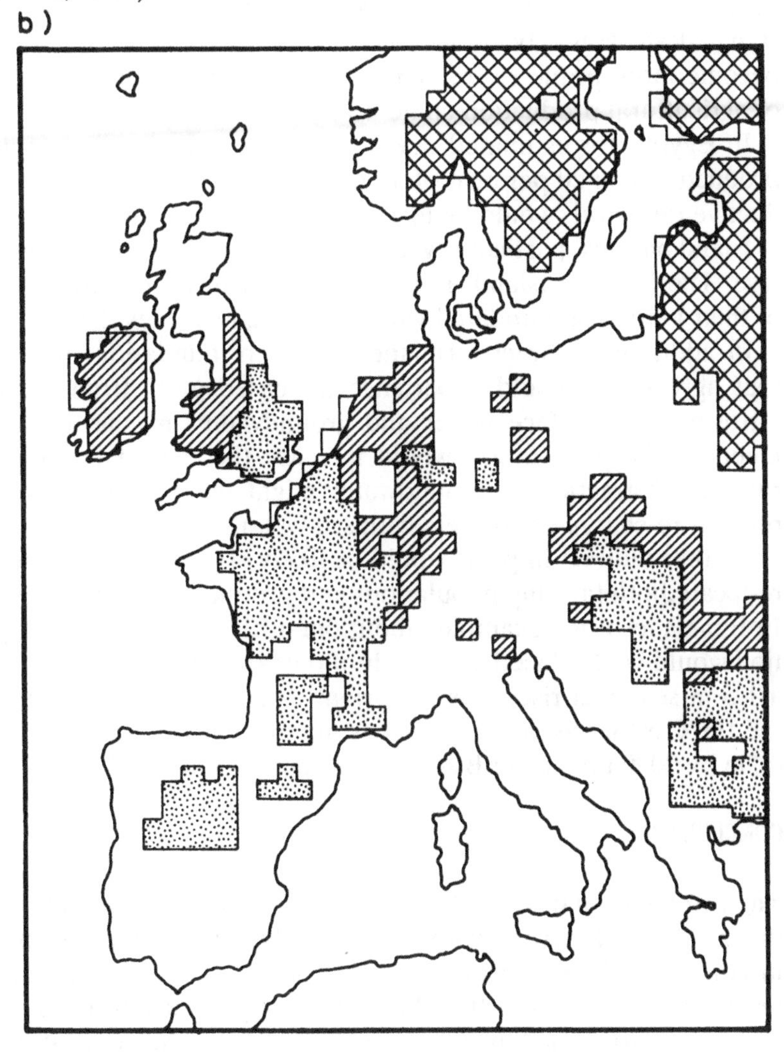

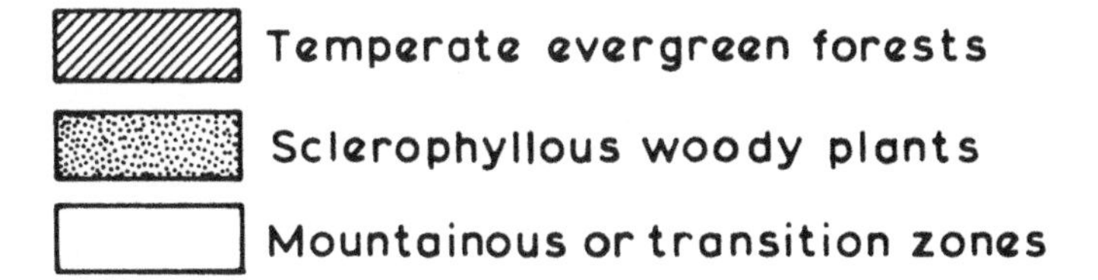

from inundation. South-east Asia would be most affected because of the extreme vulnerability of several large and heavily-populated deltaic regions. For example, with a 1.5 m sea-level rise, about 15 per cent of all land (and about one-fifth of all farmland) in Bangladesh would be inundated and a further 6 per cent would become more prone to frequent flooding.[45] Altogether 21 per cent of agricultural production could be lost.

In Egypt, it is estimated that 17 per cent of national agricultural production and 20 per cent of all farmland, especially the most productive farmland, would be lost as a result of a 1.5 m sea-level rise.

Island nations, particularly low-lying coral atolls, have the most to lose. The Maldive Islands in the Indian Ocean would have one-half of their land area inundated with a 2 m rise in sea level.[45]

In addition to direct farmland loss from inundation, it is likely that agriculture would experience increased costs from saltwater intrusion into surface water and groundwater in coastal regions. Deeper tidal penetration would increase the risk of flooding and rates of abstraction of groundwater might need to be reduced to prevent re-charge of aquifers with sea water.

Further indirect impacts would be likely as a result of the need to re-locate both farming populations and production in other regions. In Bangladesh, for example, about one-fifth of the nation's population would be displaced as a result of the farmland loss estimated for a 1.5 m sea-level rise. It is important to emphasize, however, that the IPCC estimates of sea-level rise are much lower than this (about 0.5m by 2090 under the Business-As-Usual scenario).

CONCLUSION

The combination of impacts on agriculture that could stem from the direct effects of increased atmospheric CO_2, from effects of changes in climate and, in coastal regions, from sea-level rise is likely to be extremely complex. It will certainly vary greatly from region to region and from one type of farming to another. The implications for agricultural potential are considered in the next chapter.

5. EFFECTS ON AGRICULTURAL POTENTIAL

Our efforts to assess the possible effects of climatic changes on agricultural potential have followed two broad, and complementary, approaches. One has sought to estimate the possible spatial shift in climatic resources for agriculture, and the consequent shift of land use and farming types. The other has considered possible changes in yields of crops and livestock. We shall consider these in turn.

THE SHIFT OF POTENTIAL GROWING AREAS

Shifts in mid-latitude regions

A number of studies, all of them in developed countries, have sought to identify the area over which a shift of growing potential is most likely, attempting to locate those regions most vulnerable to climatic change where changes in types of farming and in farming infrastructure would be a necessary form of adaptation.[1]

There are four steps in this approach:[2]

1. to isolate the major climatic variables that determine the spatial pattern of agricultural potential in a region;
2. to establish critical levels of these variables that match observed limits to farming types, or levels of profitability, etc.;
3. to resolve changes of climate into changes in the locations at which these critical values are achieved; and
4. to map these as a shift of isopleths to identify potential impact areas.

This approach has been used in combination with three types of scenarios of climatic change: those based on arbitrary adjustments to temperature and rainfall (i.e. synthetic scenarios), GCM-derived equilibrium $2 \times CO_2$ scenarios, and those that represent the transient response of climate to radiative forcing (see Chapter 2).

Figure 5.1 Estimations of the impacts of climatic change on the geographical extent of the US Corn Belt. (a) Simulated shift based on growing degree days (GDD in °C) during the frost-free growing season (*Source:* Newman, 1980).[4] (b) Shift for 3°C temperature increase and 8 cm precipitation increase, distributed evenly over the year (*Source:* Blasing and Solomon, 1983).[3] The solid black line indicates current location of the Corn Belt.

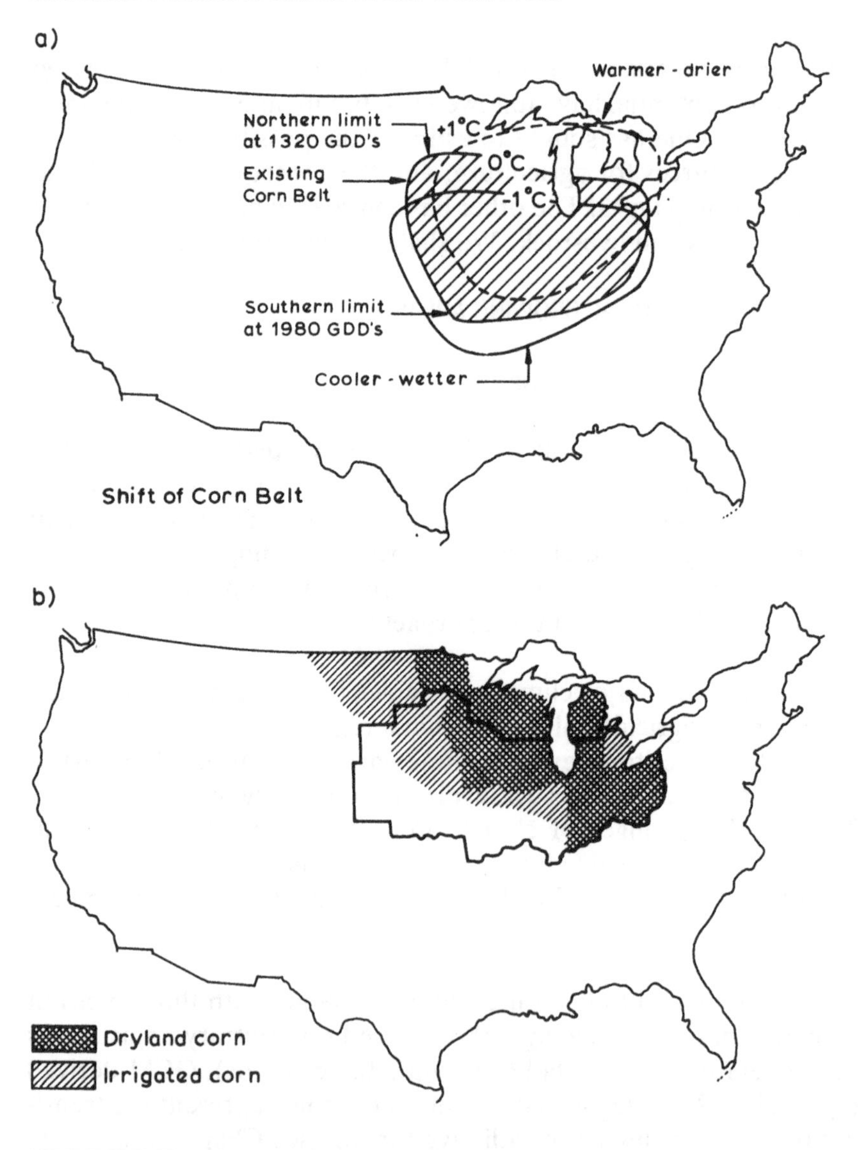

Shifts under synthetic climatic scenarios

Two separate studies, using synthetic climatic scenarios, have estimated possible shifts of agricultural potential in the Corn Belt of the central USA. Both assumed that the Corn Belt is limited in its northern extent by the length of the frost-free growing season and by the thermal requirements for maturation, and in its western extent by lack of soil moisture. They estimated the change in location of critical levels of growing-season length (measured in growing degree-days) and potential evapotranspiration for selected changes in mean temperature and precipitation. The results are given in Figure 5.1. One study found that a climatic warming would displace the Corn Belt 175 km per °C in a north-by-northeast direction.[3] The other obtained similar results though the estimated shift was less, and it concluded that if there were a concurrent increase in precipitation this could serve to counteract the displacement due to increased temperature.[4]

The shifts described here are, of course, those of agroclimatic values which currently seem to limit the US Corn Belt. Whether farmers in this region would respond by changing their cropping patterns and thus, collectively, perhaps create an actual shift of the location of corn (maize) growing would depend on many other factors, such as the competitiveness of corn against other crops. This competitiveness would almost certainly be influenced by relative changes in the yields of different crops as a result of the changes in climate, a matter considered in the second part of this chapter.

Shifts under equilibrium 2 × CO_2 climatic scenarios

More recently, impact assessments have been based on scenarios of climatic change derived from 2 × CO_2 GCM experiments. One such study, a logical development of those considered above, has mapped the shift of growing areas of different cultivars or types of the same crop that might occur under an altered climate.[5] Wheat-growing regions in North America were characterized according to their present-day temperature and rainfall regimes, and then re-mapped for the equilibrium climate based on a 2 × CO_2 experiment with the GISS GCM. Results indicated a substantial northward extension of winter wheat into Canada from its current location on the US Great Plains, a switch from hard to soft wheat in the Pacific Northwest due to increased precipitation, and an expansion of areas in autumn-sown spring wheat in the southern USA due to higher winter temperatures (Figure 5.2). In Mexico,

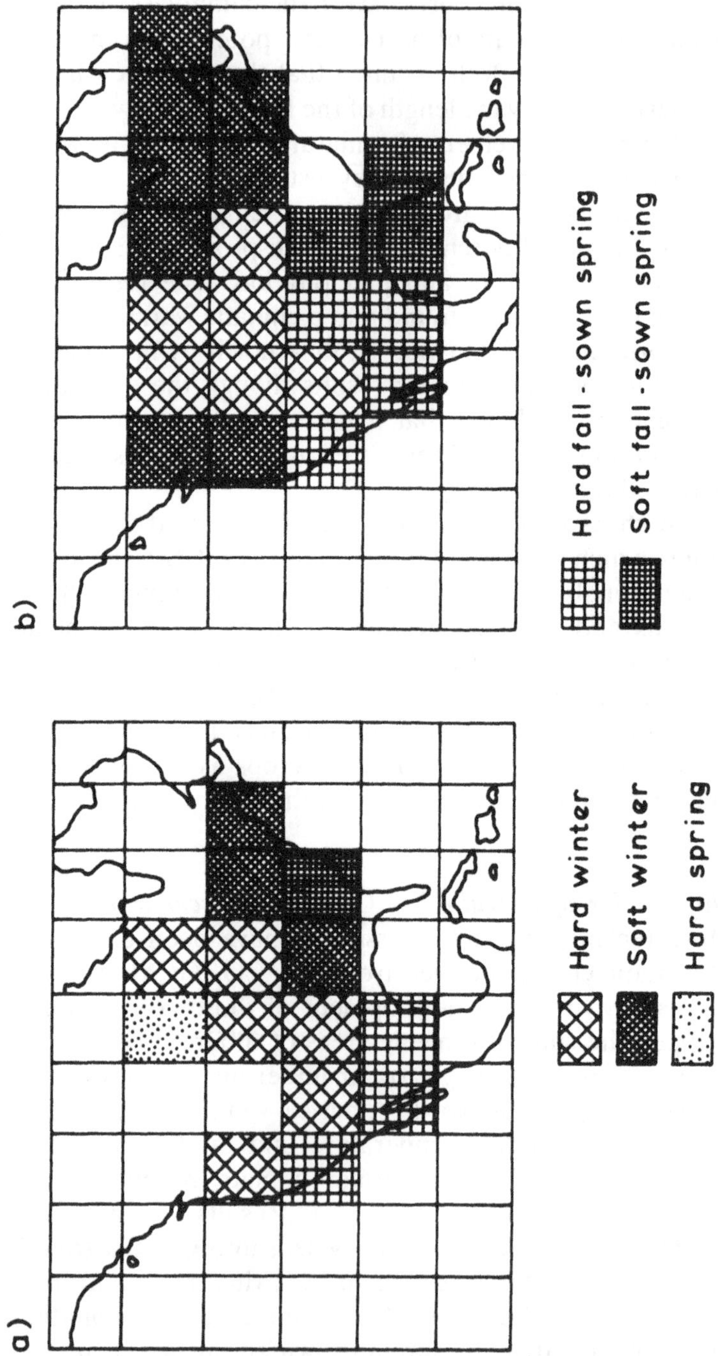

Figure 5.2 Simulated North American wheat regions using the a) GISS GCM control, and b) doubled CO_2 runs. (*Source*: Rosenzweig, 1985).[5]

wheat-growing regions would remain the same but greater high-temperature stress may occur.

A similar magnitude of shift of cropping limits has been estimated for Europe. In this region the major climatic determinant of successful ripening of grain maize (i.e. maize grown for its grain rather than as green fodder) is the warmth of the growing season. An effective temperature sum (ETS) of 850 degree-days above a base temperature of 10°C corresponds closely with the actual limit of its cultivation today.[6] This boundary extends from the south-western tip of England through northern-central Europe and central Russia to just south of Moscow. Much of the fertile north European plain is therefore currently too cool for grain maize to mature in all but the warmest years.

However, under the 2 × CO_2 equilibrium climates projected by a number of GCM experiments this limit is displaced 200 to 350 km further north. Figure 5.3 illustrates the location of the thermal limit to grain maize for 2 × CO_2 climates projected by three GCMs – GISS, GFDL and OSU (see Chapter 2). The similarity between the figures indicates the level of agreement between the models regarding temperature increases in the summer half of the year. The entire northern European plain is estimated to be within the grain maize limit under a 2 × CO_2 climate, particularly the western part of northern Europe (UK, northern Germany, Denmark) where maritime influence creates a greater sensivity to warming because greater CO_2-induced temperature increases are expected in winter than in summer. It is worth noting, however, that there is very little agreement between model estimates of precipitation, which can be a critical factor for many crops in Europe and is also important for maize. Consequently, we are at present only able to draw a very imperfect picture of how potential growing regions may shift.

Outside North America and Europe, little study has been made of the spatial shift of crop potential. One exception is Japan, where an estimate has been made of the extension of area in which rice could safely be cultivated without severe risk of crop loss due to frost (Figure 5.4). On the island of Hokkaido, in the north of Japan, the "safely cultivable" area for irrigated rice is estimated to more than double under a warming of 3.5°C (which is consistent with the GISS 2 × CO_2 climate), assuming there remains adequate precipitation and the crop is fully irrigated.[7] Corresponding poleward shifts of crop potential in the southern hemisphere, for example for cereals, fruit and vegetables in New Zealand, have also been estimated.[8] Once more, however, it must be emphasized that these are estimations

a)

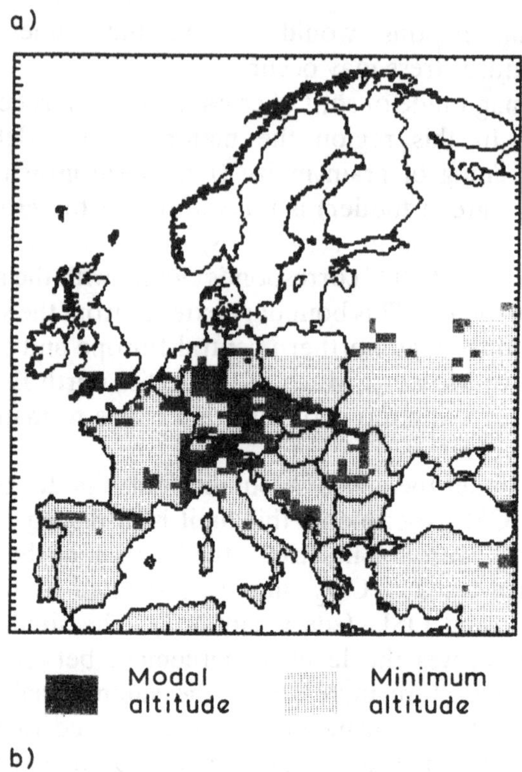

b)

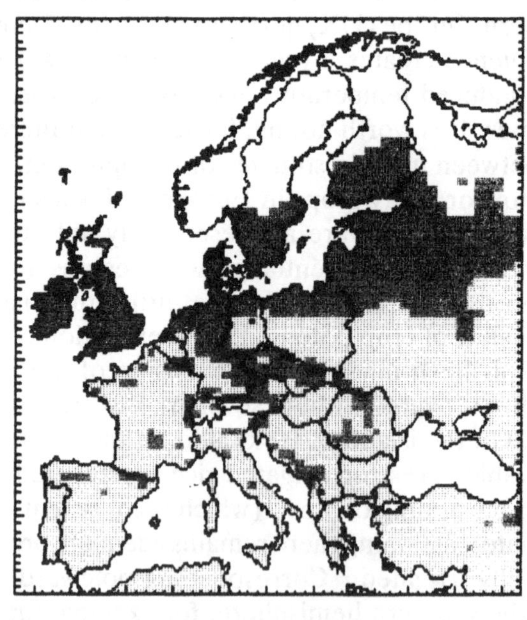

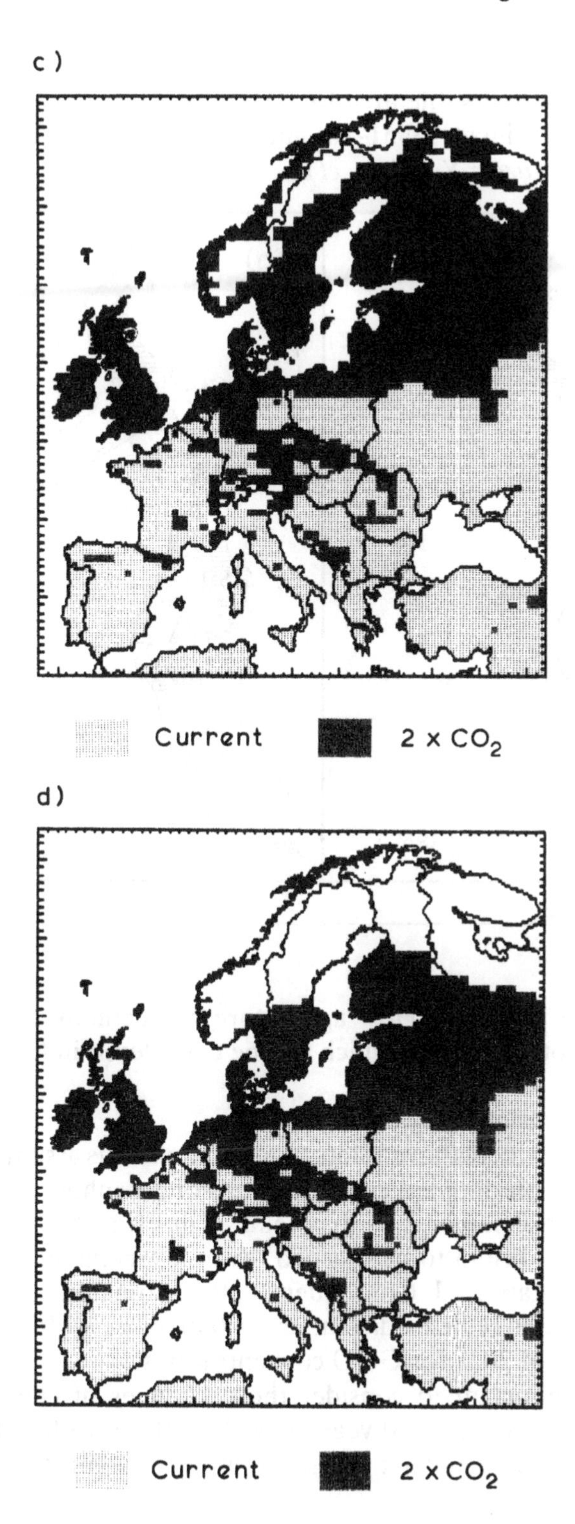

Figure 5.3 Grain-maize limits in Europe under a) current climate (1951–80), and b) GISS, c) GFDL, and d) OSU equilibrium 2 × CO_2 climates. (*Source*: Carter, *et al.*, 1990).[6]

Figure 5.4 Safely cultivable area for irrigated rice in northern Japan under a) current climate (1951–80) and b) the GISS equilibrium 2 × CO_2 climate. (*Source:* Yoshino, *et al.*, 1988).[7]

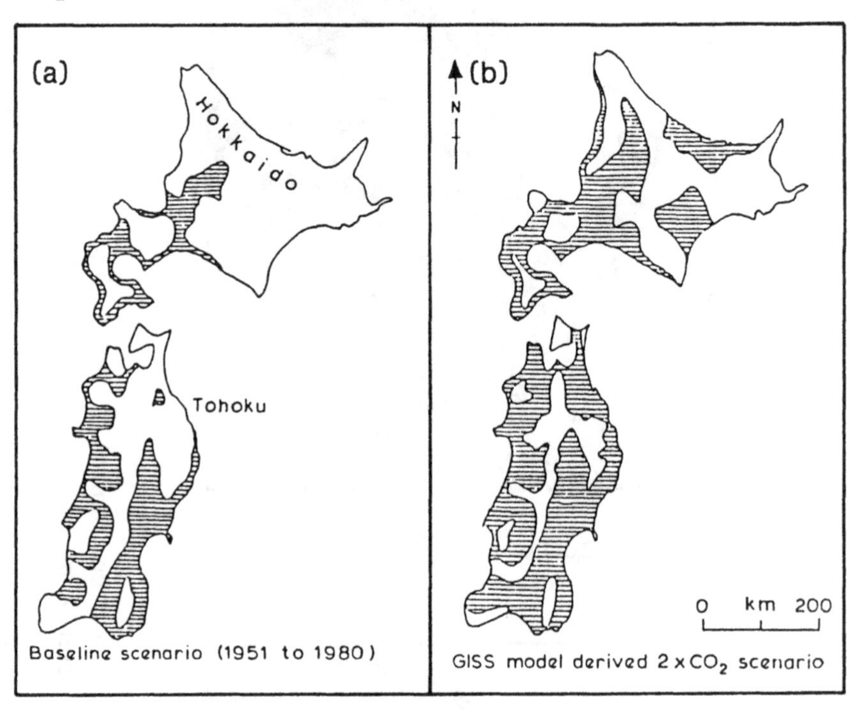

only of altered potential. How agriculture will actually respond to this is quite another matter, which will be considered in Chapter 6.

Shifts under "transient" climates

The impacts described above relate to a climate that is assumed to be in equilibrium with doubled concentrations of greenhouse gases. In fact, as noted in Chapter 2, it may take as much as half a century for such an equilibrium to be reached, even if greenhouse-gas concentrations are stabilized. Consequently, the estimated equilibrium climate response may be as distant as 2060 or more with respect to a doubling time of 2025 for GHG concentrations.

Since it is important to consider those changes of climate that may occur within the next 30 years as well as those within the next century, more recent impact assessments have begun to evaluate

Figure 5.5 Grain maize limit under the GISS transient response Scenario A in the 1990s, 2020s, and 2050s (relative to the limit for the current climate). (*Source:* Carter, *et al.*, 1990).[6]

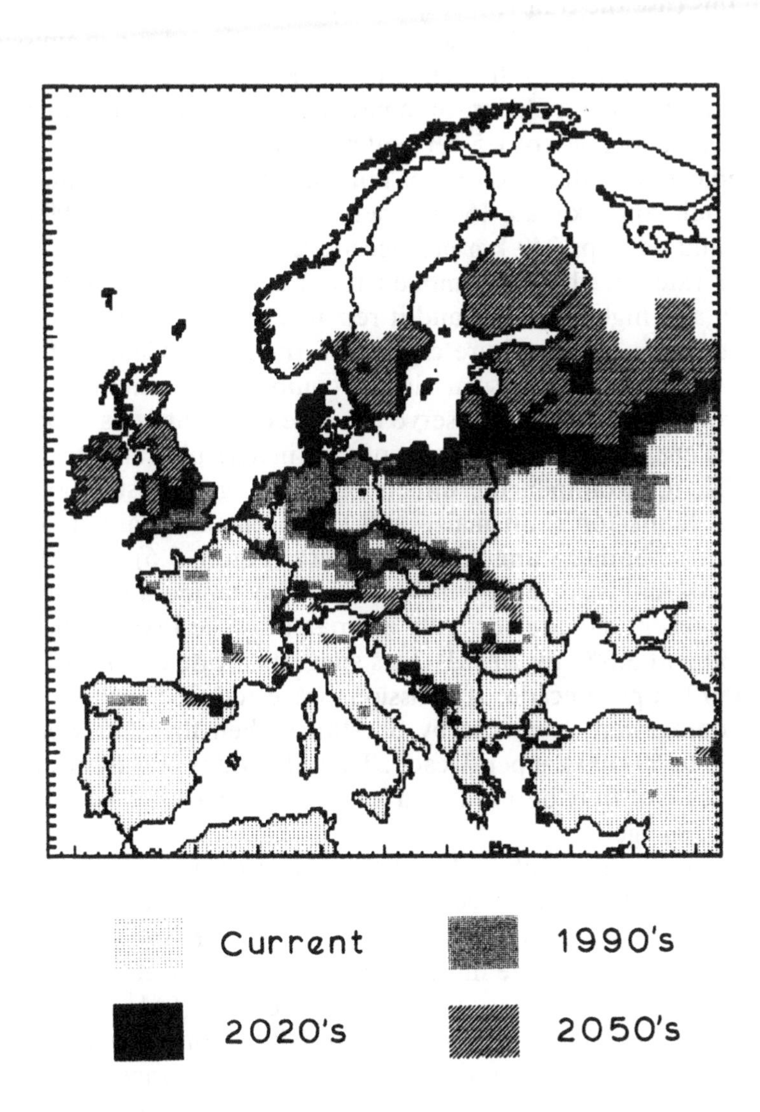

potential effects of time-dependent changes in temperature.[6]

Figure 5.5 illustrates the thermal limit for grain maize in Europe for decadal "time slices" of temperature based on results from experiments for the GISS GCM. The value of this approach is that it can help elucidate the rate of shift of agricultural potential that may occur as a result of global warming.

In this case the transient response data are for Scenario A (one of four conducted at GISS) which assumes a continued rise in emissions of trace gases at growth rates typical of the 1970s and 1980s (i.e. without effective policies of emissions control) representing an exponential increase of 1.5 per cent per annum.[9] Under this scenario the indication is that the rate of northward shift of the grain-maize limit could approximate 150 km per decade between the 1990s and 2030, and perhaps 240 km per decade from 2030 to 2060. Broadly similar rates of shift are implied for many crops throughout the middle and high latitudes, and it remains to be seen whether rates of adaptation in agriculture can match them (see Chapter 8). The use of synthetic climatic scenarios, in combination with the transient ones considered above, can serve to relate different rates of possible climatic change to different rates of shift in agricultural potential. In the UK, for example, the effects of warming suggest a poleward shift of limits for grain maize and silage maize by about 300 km for each °C in mean annual temperature (Figure 5.6).[10] Under the 'Business-As-Usual' emissions scenario the temperature increases above present day are currently estimated by the IPCC to be 1.1°C by 2030 and 3.3°C by 2090.[11] This suggests a rate of shift of about 100–150 km per decade. If emissions were reduced such that rates of warming were (say) cut by one-third, then the shift would be reduced to 50–100 km per decade. Tolerable rates of shift in climatic resources can thus be used as a guide to target rates of tolerable climatic change.

It should be remembered that the above data are for a warmer average climate. Year-to-year variations could still be expected to occur around this average, just as they do now. For example, the thermal limit for grain maize in the very warm summer of 1976 lay well north of its present average position (Figure 5.6a). Indeed, the interannual scale of shift of crop limits, from the warmest to the coolest years, is broadly similar to the shift in long-term average expected to occur under a 2–3°C warming (Figure 5.6b). Any future climate would thus have embedded in it the year-to-year variations of growing season that we experience now, but whether the range of these variations will be similar to the range experienced today is uncertain.

Figure 5.6 Hypothetical limits for successful ripening of two crops based on temperature: (a) Grain-maize (requirement: 850 degree-days above a base temperature of 10°C), and (b) Silage maize (requirement: 1,460 degree-days above a base temperature of 6°C). Mean limits (thick solid lines) are representative of lowland conditions, based on temperature data from 78 stations for the period 1951–80. Also shown are limits for individual years (open circles) and limits for arbitrary adjustments in mean temperature (broken lines). (*Source:* Parry, *et al.*, 1989).[10]

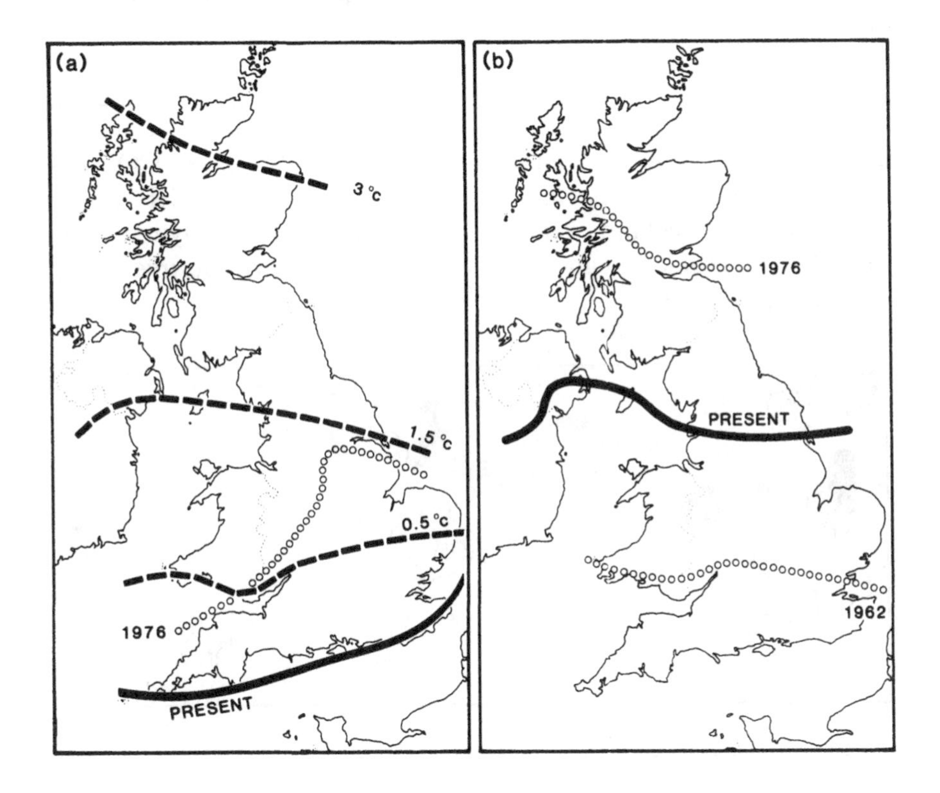

The shifts of crop potential described above are examples of a worldwide relocation of climatic zones that could occur as a result of CO_2-induced changes of climate, particularly a poleward shift of thermal zones. An illustration of the scale of these shifts is given in Figure 5.7 which maps regions that have a present-day climate analogous to the future climate assumed under the GISS 2 × CO_2 scenario. For example, Iceland's climate estimated under the GISS 2 × CO_2 scenario is similar to that of northern Britain today. This serves to illustrate not only the magnitude of possible changes in agricultural potential, but also the adaptive responses likely to be required to re-tune agriculture to altered climatic resources.[12] For example, perhaps the combination of barley growing and cattle rearing and fattening, which are successful enterprises in northern Britain today, would be appropriate for Iceland in the future. Due to differences in latitude, however, there are important differences in day length between such regions, and the analogy is far from perfect.

Figure 5.7 Present-day analogues of the GISS 2 × CO_2 climate estimated for selected regions in the IIASA/UNEP study: Saskatchewan, Iceland, Finland, Leningrad and Cherdyn regions (USSR) and Hokkaido and Tohoku districts (Japan). (*Source:* Parry & Carter, 1988).[12]

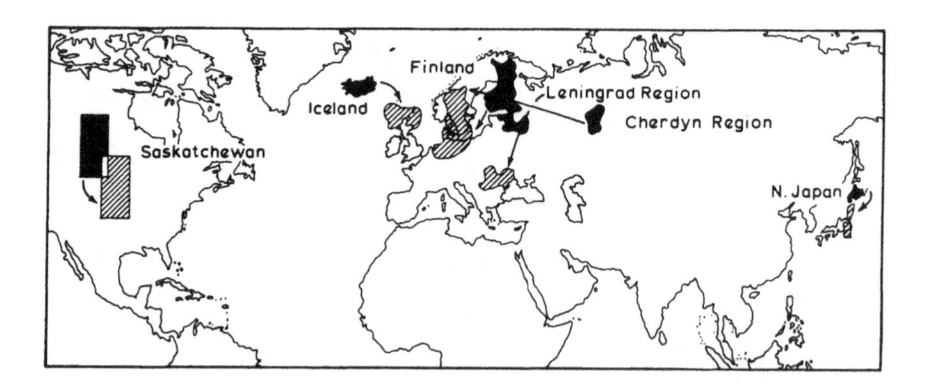

Shifts in altitudinal limits

Increased temperatures can also be expected to raise altitudinal limits to cultivation where these are currently determined by low average levels of warmth and high risk of frost. An illustration of this is provided by studies of changing levels of risk on upland farming in the British Isles. It has been estimated that a critically high rate of crop failure due to cold summers is one of the most important factors affecting the location of the upper limit of cereal cultivation in this region.[13] The maximum tolerable frequency can

Figure 5.8 Shift of 1 in 3.3 failure frequency for oats in the British Isles for 1°C increase in mean annual temperatures (normals 1856–95). (*Source:* Parry, 1985).[2]

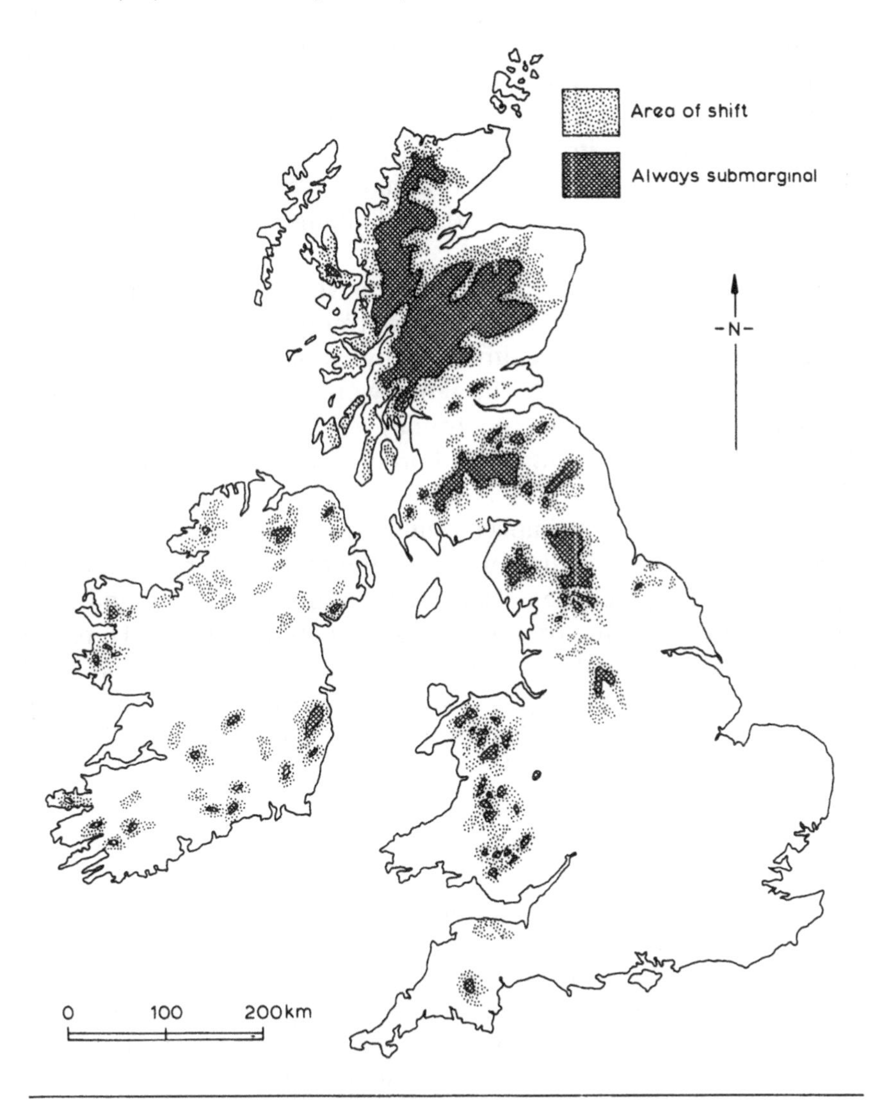

certainly be no higher than one year in three and, when mapped for the British Isles, the isopleth of this frequency delimits a region above which, under current climate, cereal cropping is too risky to be worthwhile.[14] This submarginal area, covering about six million ha, lies above 500 m in southern Britain but is found as low as 400 m in central Scotland due to the lower mean temperature at higher latitude.

An increase of 1°C in mean annual temperatures, which could occur by about 2030 in Britain under the Business-As-Usual scenario,[11] is estimated to lead to a 140 m upward shift of the isopleths of risk of crop failure, assuming an unaltered probability distribution of cold summers around the new mean (Figure 5.8).[2] In total about one-third of Britain's unimproved moorland, which is at present submarginal for cereal farming, would become marginally viable in terms of its summer warmth. Other constraints such as steep terrain and acid soils would, however, remain, so we should expect the response to this change in potential to be somewhat muted.

More recent studies in the European Alps reveal a similar scale of potential impact. Here a 1°C warming can be expected to raise climatic limits of cultivation by about 150 m and a 4°C warming by 450–650 m.[15] The latter would imply a raising of the climatic zones of the Alps to altitudes similar to those today in the Pyrenees on the Franco-Spanish border which lie 300 km south of the Alps.[15]

Possibly the greatest impact would be felt in the Andes, where altitude rather than terrain and soils is the major determinant of the upper limit of agriculture because rich basaltic soils are found even under the high paramo or alpine grassland above 4000 m. Here the risk of frost to crops of beans, barley and potatoes is closely related to mean winter temperatures, and a 1°C warming would probably raise climatic limits of cultivation by about 200 m (from 3800 m to 4000 m in central Ecuador).[16]

Shifts in low-latitude regions

While effects of greenhouse gas-induced changes in temperature on the location of growing potential may be less at low than at middle and high latitudes, due to the smaller increases in temperature expected here, the effects of any changes in precipitation could be substantial. Unfortunately, however, no comprehensive study has yet been made of the impact that climatic change could have on the location of agricultural potential in the tropics. The following is a survey of the scattered and anecdotal evidence that is at present available.

More northerly penetration of the Intertropical Convergence Zone (ITCZ) should, on average, increase summer rainfall receipt in the Sahel, but average levels of soil moisture may decrease as a result of higher rates of evaporation due to higher temperatures (see Chapter 2, above). An indication of the possible impact of this on the location of agricultural potential is given by the effects of drought in the region in the 1970s. In Senegal, for example, annual rainfall receipts over the 16 years 1968–83 were 30–35 per cent less

Figure 5.9 Percentage change in net primary productivity relative to the present climate for a climate scenario roughly equivalent to a doubling of atmospheric carbon dioxide (for details, see text). (*Source:* Pittock & Nix, 1986).[18]

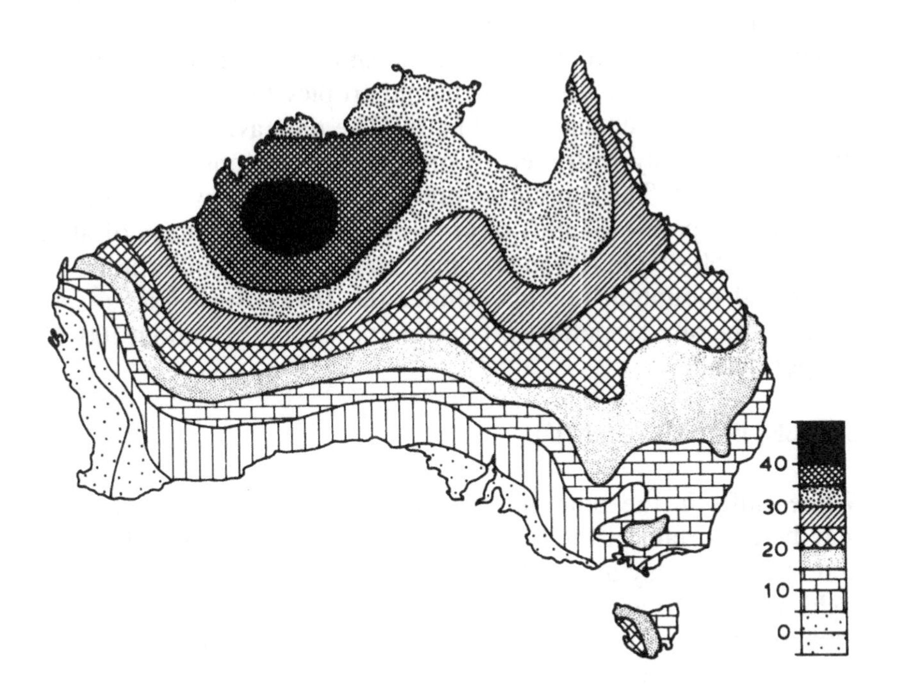

than the 50-year mean, and over the 30 years 1954–83 were 15–20 per cent less.[17] During the latter 30-year dry period the rainfall isophets would be drawn 60 km further south than the 50-year average, and during the 16-year dry period more than 110 km further south. On the basis of this experience it seems that agroclimatic zones in Senegal could shift about 20–30 km per 10 per cent increase or decrease in mean annual rainfall. A similar scale of change is indicated for semi-arid regions in Australia. Figure 5.9 shows the estimated change in potential biomass production relative to the present climate for a climatic scenario that assumes increases in mean annual temperature of 0.1°C for every degree of latitude (i.e. by 1°C at 10°S and 4°C at 40°S), together with 40 per cent increases in summer precipitation, and 20 per cent decreases in winter precipitation.[18] Approximately half of Australia experiences an increase in potential biomass production of over 20 per cent. The only decrease is in the south-west of Western Australia, an area with almost no summer rainfall so that the decrease in winter rainfall is the dominating factor. It should be noted that no account is taken here of other climatic elements such as solar radiation, humidity and windspeed, nor of the possible direct effects of increased ambient CO_2 concentrations (see Chapter 4).

Much more study is needed of the shift of zones of agricultural potential in the semi-arid and humid tropics that may occur as a result of possible changes of climate. The data available at present are almost wholly restricted to mid-latitude regions and to developed countries. These indicate that a 1°C warming would induce a 200–400 km poleward shift of cropping zones at latitudes of about 50° and a 150–200 m rise in cropping zones in both mid-latitude and equatorial mountain regions. Under the IPCC Business-As-Usual scenario the rate of spatial shift of climatic resources is about 100–150 km per decade.

CHANGES IN YIELD POTENTIAL

While shifts in the potential limits for different types of farming may be important at the boundaries of current agricultural regions and in present-day areas of marginal farming, changes in potential yield in the core areas of today's main food-producing regions will probably have a greater impact on overall production. The degree of this change will vary from region to region and crop to crop, with quite complicated resulting patterns of impact.

Regional differences in impact

Not only are there likely to occur varying degrees of absolute change in climatic variables (such as greater warming at higher latitudes), but the effect of these changes will be very much a function of the change in climate relative to existing conditions. To illustrate, while summer warmth in southern Finland may increase by one-third under the GISS 2 × CO_2 scenario, in northern Finland (where it is already one-tenth lower than in the south), the increase is over 40 per cent.[19] Both the absolute and relative warming is thus greater at higher latitudes and the effects of this can be expected to be widespread. As a result, the estimated increases in rice yields under a 2 × CO_2 climate in Hokkaido in the north of Japan are about twice that in central Japan (Tohoku).[7] Similar differential effects between latitudes are found elsewhere and we shall see later that this may have profound implications for the balance of advantage between higher-latitude and lower-latitude regions in terms of their agricultural potential.

Non-linear effects

We saw in Chapter 4 that changes in temperature and precipitation may have a non-linear effect on crop yields, and that different crops can respond quite differently to such changes. These differences in response could have a major effect on the future use of land.

Consider, for example, the yield response to increases in average temperature of barley and wheat in the Moscow region of the USSR shown schematically in Figure 5.10. Here wheat is currently near its northern limit of ripening and yields of barley are on average higher and less variable. But beyond a certain amount of warming barley becomes heat-stressed in particularly warm summers while wheat yields increase. If we assume that the crops are otherwise equally competitive, the cross-over of the curves in Figure 5.10 indicates the point at which the two crops are equally profitable. To the left of the cross-over it would make sense to grow barley, and to the right wheat. In this simple example a small climatic change could induce a major change in choice of crop and thus of the land use of the region. In reality patterns of cropping and livestock production throughout the world are the result of the intersection of similar (though much more complex) functions. A small change in one of these functions, due for example to a change in climate, could bring about a radical change in regional production patterns.

Figure 5.10 Hypothetical yield-temperature response curves for two crops (A and B) in the same region. (*Source:* Parry & Carter, 1988).[12]

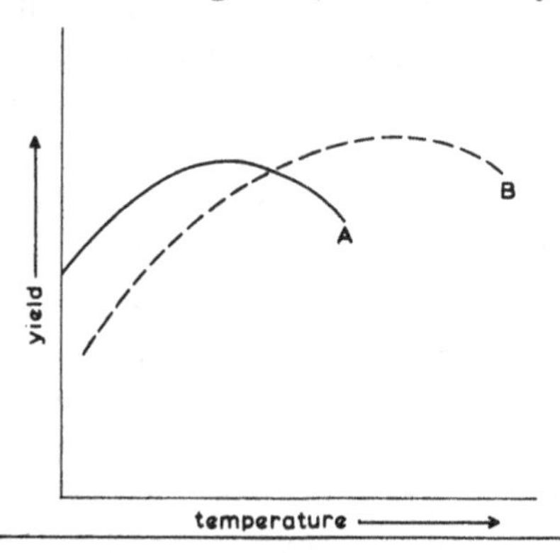

Effects on the yield potential of staple cereal crops

Northern regions

In the high mid-latitudes of the northern hemisphere the series of case studies by IIASA and UNEP indicate quite substantial increases in productive potential where warming is expected to reduce the current constraints imposed by inadequate temperature. A summary of estimated impacts is given in Figure 5.11.

Agriculture in Scandinavia stands to gain more from global warming than perhaps any other region of the world. For example, in Finland, where the equilibrium 2 × CO_2 climate is projected to be about 4°C warmer and also wetter than at present, yields of adapted cultivars of spring wheat are estimated to increase by about 10 per cent in the south, up to 20 per cent in the centre and even more in the north. Yields of barley and oats are raised by 9–18 per cent, depending on the region in Finland.[19]

In northern Japan, where temperature is projected to increase by 3–3.5°C and precipitation by 5 per cent (the GISS 2 × CO_2 climate),

Figure 5.11 (opposite page) Estimated crop yields under the GISS 2 × CO_2 scenario for present-day and for adjusted crop varieties and management Finland, N. USSR and N. Japan. 1 = present variety; 2, 3, 4 = varieties with thermal requirements 50, 100 and 120 GDD higher than present; 5, 6= newly introduced middle-maturing and late-maturing rice varieties, transplanting date 25 days earlier than present; 7 = present variety with technology trend projected to 2035;8 = present variety, fertilizer applications 50% above those in 7;9 = present variety, drainage activity 2 km/km^2 above that in 7; 10 = combination of 8 and 9; 11 = includes “direct” effects of CO_2. (*Source:* Parry & Carter, 1988).[12]

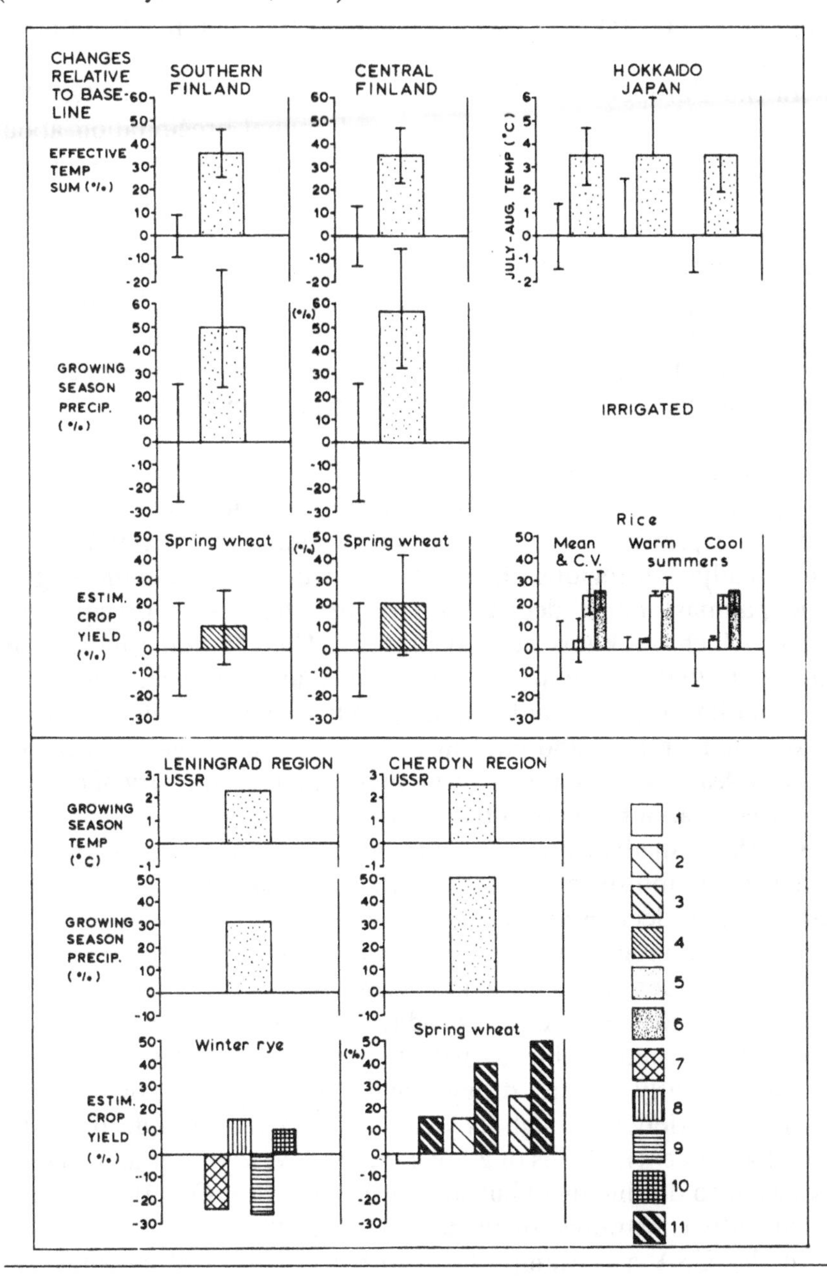

rice yields are estimated to increase in the north (Hokkaido) by about 5 per cent, and in the north-central region (Tohoku) by about 2 per cent.[7] The average increase estimated for the country as a whole is about 3 per cent. Cultivation limits for rice would rise approximately 500 m in elevation and advance about 100 km north in Hokkaido.

In the European USSR (Leningrad region), with May–October temperatures 2–3 degrees warmer and annual precipitation about 100 mm higher (the GISS 2 × CO_2 climate), yields of winter wheat and maize are likely to increase but those of temperate crops such as barley, oats, potatoes and green vegetables are likely to decrease.[20] In the Perm region, just to the west of the Ural mountains at about 60°N, spring-wheat yields are expected to decrease slightly under the warmer growing season, but this may be more than compensated for by the direct effects of CO_2, the combined climate and direct CO_2 effects perhaps allowing a 20 per cent increase in yields.[20]

Mid-latitude grain belts

In today's breadbasket regions of the world (the US Great Plains, Canadian prairies, North European lowlands, the Soviet Ukraine and its adjacent regions, the Australian wheat belt, and the Argentine pampas) much depends on future changes in precipitation about which we know little at present. There is some indication, however, that less moisture may be available for plant growth in the mid-latitude mid-continental regions (which include the Great Plains and prairies and current grain producing regions in Soviet central Asia – see Chapter 2). The following is a summary of current knowledge about effects in these areas.

On the Canadian prairies a warming of 3–4°C, accompanied by reduced soil moisture consistent with the GISS 2 × CO_2 climate, is estimated to decrease yields of spring wheat nationally by about 19 per cent, with regional variations from 18 per cent in Saskatchewan, to about 10 per cent in Manitoba, and with a small increase near the current northern limit of production.[21, 22] Winter wheat would probably be better able to withstand an increased frequency of spring and early summer drought and might expand at the expense of spring varieties (though its yield is also expected to decline – by about 4 per cent). Yields of grain corn, barley, soybeans and hay are expected to decline in all but the northern part of Ontario, where it is currently constrained by inadequate warmth.[22]

In the USA a warming of 3.8 to 6.3°C, with soil moisture reduced by 10 per cent (which is consistent with the GISS and GFDL

Figure 5.12 Estimated maize yields in the USA under the GISS and GFDL 2 × CO_2 climates with and without the direct effects (DE) of CO_2: a) dryland and b) irrigated. Estimations for direct effects assume CO_2 concentrations of 660 ppm, which are 100 ppm above IPCC estimates and are thus somewhat exaggerated (*Source:* Rosenzweig, 1989.)[23]

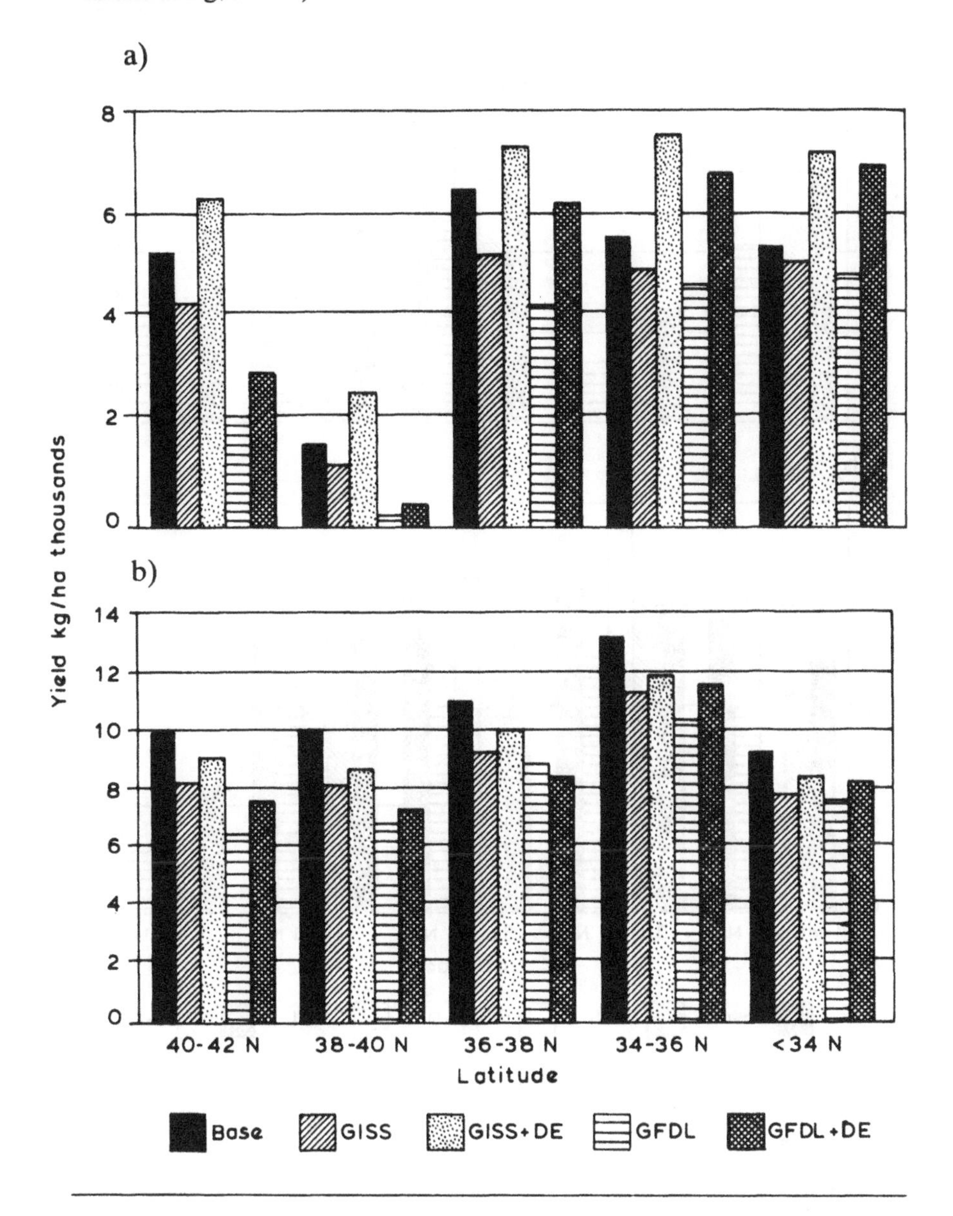

Figure 5.13 Estimated wheat yields in the USA under the GISS and GFDL 2 × CO_2 climates with and without the direct effects (DE) of CO_2: a) dryland and b) irrigated. (*Source:* Rosenzweig, 1989).[23]

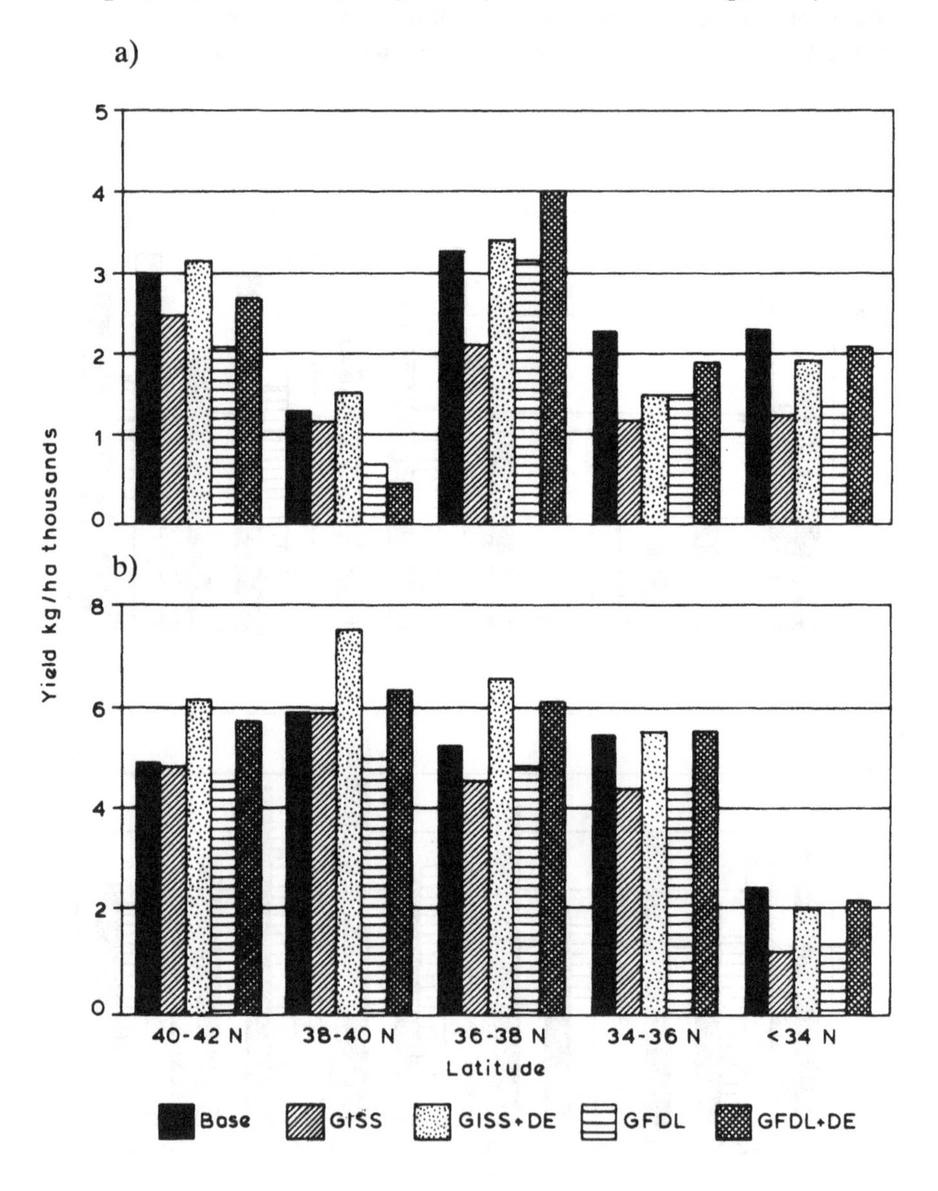

$2 \times CO_2$ climate), is estimated to lead to a decrease in potential yields of maize allowing for the limited beneficial fertilizing effect of enhanced CO_2 on this C4 plant. The decrease would be 4–17 per cent in California, 16–25 per cent on the Great Plains (assuming irrigation), and 5–14 per cent in the south-east (also assuming irrigation) (Figure 5.12). In the Great Lakes region there could be a small increase in potential yields, depending on available moisture.

Under the same changes of climate, potential wheat yields in the USA are estimated to decrease by 2–3 per cent, and irrigated yields increase by 5–15 per cent, these increases due largely to the projected beneficial effects of more atmospheric CO_2 (Figure 5.13). Dryland soybean yields show a wide range of decreases from −3 per cent in the Great Lakes region to between −24 to −72 per cent in the south-east. Irrigation could offset much of these losses.[23]

In northern Europe, where mean annual temperature increases of 3.5–4.5°C are projected for a $2 \times CO_2$ climate, the response of yields of wheat, maize and other cereals are likely to depend very much on corresponding changes in available moisture. If precipitation increases in both winter and summer, present moisture levels are likely to be maintained. But if summer rainfall decreases or is unaltered while winter rainfall increases, then an increase in irrigation would be necessary to maintain potential yield levels. In this case current yield levels of winter wheat could probably be maintained.[24, 25]

In southern Europe quite substantial decreases in productive potential could occur if the GCMs are correct in predicting decreases in soil moisture in the summer, and possibly also in the winter months. Under a warming of 4°C and with summer rainfall reduced by 15 per cent (the UKMO $2 \times CO_2$ climate) biomass potential is estimated to decrease in Italy and Greece by 5 per cent and 36 per cent, respectively.[25] In general there is a quite striking contrast between the increases in productive potential in northern Europe and the decreases in southern Europe that are suggested by current GCM experiments for a doubling of GHG. This implies an important northward shift of the balance of agricultural resources in the European Community.

In the USSR grain belt the GISS $2 \times CO_2$ scenario implies an increase in temperature and precipitation during the growing season of 3.3°C and 22 per cent, respectively. As a result, spring-wheat yields in the Saratov region in the south-east Ukraine could increase by 10–15 per cent, but would decrease significantly if precipitation did not increase.[20] Recent studies by Soviet scientists have used palaeoclimatic analogues as a basis for assessing possible impacts of future climatic change. The Holocene Optimum is taken as an analogue for the year 2000 (+ 1°C, with a CO_2 concentration of 380 ppm), the Eemian Interglacial for 2030 (+ 2°C, 420 ppm) and the Pliocene Optimum for about 2050 (+3 to +4°C, 560 ppm). These studies suggest that moisture levels in drier parts of the mid-latitudes might decrease in the initial part of the warming phase, with consequent decreases in potential productivity. However, the Eemian and Pliocene palaeo-analogues suggest an increase in moisture availability which, combined with the beneficial direct effects of increased atmospheric CO_2, would probably lead to increased potential productivity in agriculture. At present, however, it is not clear how far it is appropriate to adopt earlier warm epochs as analogues of possible future climatic change.[20]

In Australia, there are indications that increasing GHG will lead to increases in summer rainfall where there is at present a summer rainfall maximum (e.g. the eastern wheat belt) and decreases in winter rainfall where there is at present a winter rainfall maximum (e.g. the western wheat areas). This could increase potential yields in the east, and decrease them in the west.[26]

Information concerning possible impacts in other grain-exporting regions is extremely limited. There is some suggestion that higher temperatures (about 2–4°C) for doubled GHG could reduce available moisture in the cereal-farming areas of southern Africa.[27] Even without this information, however, it is reasonable to conclude that while there may be important increases in productive potential in some major grain-producing regions in the mid-latitudes, the overall picture is one of reduced potential.

Semi-arid tropics

There is enormous uncertainty here, stemming from our ignorance of future changes in precipitation – both its quantity and its distribution over time and space. Increases in temperature, even the relatively small increases of about 1.5°C projected for lower latitudes for a doubling of GHG, would increase rates of evapotranspiration by 5 to 15 per cent and this, if there were not compensating increases in rainfall, would tend to reduce yields. In northern India temperature increases would tend to reduce wheat yields due to heat stress, possibly by about 10 per cent for a 0.5°C warming if rainfall did not increase. However, rice yields might increase under higher temperatures if rainfall increased.[28]

A similar increase in yields could occur in China if global warming led to a weaker winter but stronger summer monsoon. Rainfall receipt would thus increase in the already rainy areas but extend further westward and northward than at present. If rainfall consequently increased by about 100 mm, a 1°C warming might lead to increased yields of wheat, rice and maize by about 10 per cent.[29] Without increases in available moisture, however, maize yields in the eastern and central regions could decrease on average by 3 per cent per 1°C.[30]

No impact studies based on climatic scenarios from GCM experiments have, at the time of writing, been made in Africa, primarily because there is little agreement between GCMs concerning changes in rainfall, which is the climatic variable that most determines variations in agricultural yield in this region. Some preliminary work has been completed in South Africa, indicating that an increase in mean annual temperatures of 2°C to 4°C in Natal Province (consistent with the range of GCM projections for 2 × CO_2 climates) would increase rates of evapotranspiration by 5 per cent to 15 per cent and reduce biomass productivity by about the same amount.[27]

In Kenya, a case study of the IIASA/UNEP project considered the effects of the driest 10 per cent of years occurring at present. These suggest reductions of maize yields by 30 per cent–70 per cent.[31] Any changes in the frequency of such years would substantially affect the average output of agriculture in the region, but we cannot at present estimate how these might change.

Humid tropics

Intensification of the south-east Asian monsoon would tend to lead to increased summer rainfall but also possibly reduced winter rainfall, which would in turn affect the amounts of water available for wet rice (paddy) production. Preliminary results from a UNEP-funded project indicate that in northern Thailand, with rainfall changes of +5 per cent (summer) and −11 per cent (winter), irrigation requirements for rice would decrease by about 3 per cent in summer and increase by about 30 per cent in winter.[32] Resulting potential rice yields would decrease by 1 – 7 per cent, although the beneficial direct CO_2 effect could enable increases of 4 per cent–13 per cent. If, however, increases in temperature are also considered, the more rapid growth of the crop can be expected to reduce yields overall and to increase losses to pests.

A great deal of uncertainty surrounds the possible effect of climatic changes on potential rice yields in the humid tropical and equatorial regions, primarily because yields are extremely sensitive to the amount of water available at certain times in the growing season, because of our relative ignorance of how rainfall may alter in its amount and its timing, and because of uncertainty about how rice will respond to elevated levels of CO_2 in the field (rather than in experimental glasshouses). Most of this uncertainty will take years, if not decades, to resolve.

Effects on grassland yield and livestock carrying capacity

Here the information is extremely scant. The only model-based study using GCM-derived scenarios of future climate has been completed in Iceland where increases in mean annual temperature of 4°C and in rainfall of 15 per cent (consistent with the GISS $2 \times CO_2$ climate) are estimated to increase the carrying capacity of sheep by improved grassland by about two-and-a-half times, and by rough pasture by more than a half.[33]

Few other regions, however, are currently as constrained by inadequate warmth as Iceland and thus are unlikely to benefit as much from global warming. The exceptions might be in Patagonia, in the southern part of Argentina and Chile, where grass production and cattle grazing are limited by temperature rather than rainfall. Further north, in the southern pampas of Argentina, increases in rainfall would be needed to compensate for higher rates of evapotranspiration (about 10 per cent) that would stem from higher mean temperatures (2°C to 4°C under the GCM $2 \times CO_2$ climates).

The productivity of the rangelands of Africa depends almost wholly on the amount and timing of rainfall. In Kenya, for example, forage yield in the driest 10 per cent of years is reduced by 15 per cent to 60 per cent from its average.[31] The carrying capacity of livestock can thus fall by 10 to 40 per cent and milk yields to zero.

Projected increases in summer rainfall in eastern Australia are expected to increase grass growth but this is likely to be offset by the poorer nutritive value of tropical compared with temperate species. Loss of the Mediterranean-type climatic zones of Victoria and Western Australia, which are the current principal lamb and wool producing areas, together with increased heat stress of both cattle and sheep, could mean that livestock productivity would decrease.[34]

CONCLUSIONS

Impacts on potential yields vary greatly according to types of climatic change and types of agriculture. In general, there is much uncertainty about how agricultural potential may be affected.

In the northern mid-latitudes where summer drying may reduce productive potential (e.g. in the US Great Plains and Corn Belt, Canadian prairies, southern Europe, south European USSR) yield potential could be reduced by 10–30 per cent under an equilibrium $2 \times CO_2$ climate. However, towards the poleward edge of current core-producing regions (e.g. the northern edge of the Canadian prairies, northern Europe, northern USSR and Japan, southern Chile and Argentina) warming may enhance productive potential, particularly when combined with beneficial direct CO_2 effects. Much of this potential may not, however, be exploitable owing to limits placed by inappropriate soils and difficult terrain, and on balance it seems that the advantages of warming at higher latitudes would not compensate for reduced potential in current major cereal-producing regions.

Effects at lower latitudes are much more difficult to estimate because production potential is largely a function of the amount and distribution of precipitation and because there is little agreement about how rainfall may be affected by GHG warming. Because of these uncertainties the tendency has been to assert that worthwhile study must await improved projection of changes in precipitation. Consequently very few estimates are currently available of how yields might respond to a range of possible changes of climate in low-latitude regions. The only comprehensive national estimates available are for Australia and New Zealand where increases in cereal productivity might occur (except in western Australia) if warming is accompanied by an increase in summer rainfall.[35, 8]

The impacts described above relate to possible changes in potential productivity or yield. It should be emphasized that such potential effects are those estimated assuming present-day management and technology. They are not the estimated future actual effects, which will depend on how farmers and governments respond to altered potential through changes in management and technology. The likely effects on actual agricultural output and on other measures of economic performance such as profitability and employment levels are considered in the following chapter.

6. EFFECTS ON PRODUCTION AND LAND USE

To date (1990) six national case studies have been made of the potential impact of climatic changes on agricultural production.Five were conducted by IIASA and UNEP between 1983 and 1986 in Iceland, Finland, the USSR and Japan, and in one province of Canada (Saskatchewan).[1] Since 1986 additional impact assessments have also been completed for four other Canadian regions (Manitoba, Ontario, Alberta and the Maritime provinces).[2] The most recent national study has been made in the United States by the US Environmental Protection Agency.[3] These studies are based on results from model experiments of yield responses to altered climate and the effects that altered yields might have on production. They adopted a study method similar to that adopted and tested in the IIASA/UNEP project (see Chapter 3).

Other countries have conducted national reviews of effects of climatic change, basing these on existing knowledge rather than on new research. The most comprehensive of these are for Australia and New Zealand.[4, 5] Brief surveys have also been completed in the UK and West Germany.[6, 7] Several other national assessments are currently in progress but not yet complete.

This chapter provides a summary of results from the most detailed of these surveys: the model-based studies completed for six regions, and the comprehensive national reviews for Australia and New Zealand. These provide us with an array of assessments for three world regions: the northern and southern mid-latitude grain belts and northern regions at the current margin of the grain belt. We shall take these regions in turn. Unless otherwise stated the estimated effects are for climates described by 2 × CO_2 GCM experiments. No national assessments have been completed using climates described by transient response GCM experiments. Some of the estimates relate to the effect only of altered climate, others to the combined effect of altered climate and the direct effect of increased atmospheric CO_2.

EFFECTS ON PRODUCTION IN THE NORTHERN AND SOUTHERN MID-LATITUDE GRAIN BELTS

United States

A recent and comprehensive study by the EPA suggests that, in most parts of the USA, increased temperatures and reduced crop-water availability projected under the GISS and GDFL 2 × CO_2 climate experiments (+ 3.8 to + 6.3°C, soil moisture −10 per cent) would lead to a decrease of yields of all the major unirrigated crops.[3] The largest reductions are projected for the south and south-east. In the most northern areas, however, where temperature is currently a constraint on growth, yields of unirrigated maize and soybeans could increase as higher temperatures increase the length of the available growing season. When the direct effects of increased CO_2 are considered, it is evident that yields may increase more generally in northern areas but still decrease in the south where problems of heat stress increase and where rainfall may decrease (see Chapter 5).

These estimates of altered yields were used as inputs to a suite of farm-level models, subsets of a national agricultural model that represents production, consumption and land use in the USA. Experiments with this model indicate that production of most crops is reduced because of yield decreases and limited availability of suitable land. The largest reductions are in sorghum (−20 per cent), corn (−13 per cent) and rice (−11 per cent), with an estimated fall in net value of agricultural output of $33 billion. As a consequence, consumers would face slightly higher prices, although supplies are estimated to meet current and projected demand. However, exports of agricultural commodities are estimated to decline by up to 70 per cent, which could have a substantial effect on the pattern of world food trade (see Chapter 7).

With the relative increase in productivity in the north and decrease in the south of the USA, quite major northward shifts of land use are suggested, particularly in the production of wheat, maize and soybean. Crop acreage in Appalachia, the south-east and the southern Great Plains could decrease by 5–25 per cent, while acreage in the northern Great Lakes region, the northern Great Plains and the Pacific Northwest could increase by 5–17 per cent (Figure 6.1).

The effects described above are, of course, based on conjecture. They would depend on how food production is concurrently affected in other parts of the world (which would influence the world price of

Figure 6.1 Estimated changes (%) in land use, by region, in the USA in response to changes in crop yields under the GISS and GFDL 2 × CO_2 climates. (*Source*: EPA, 1989).[3]

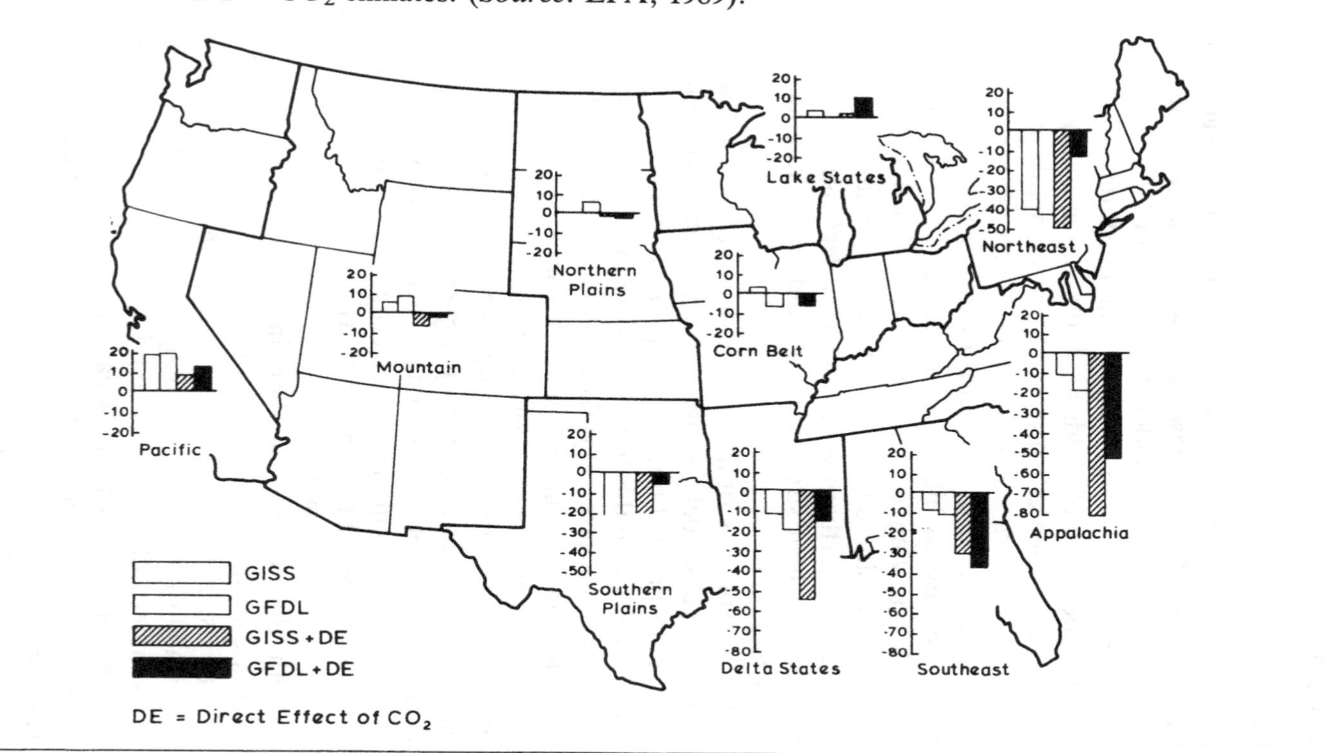

commodities, and thus prices in the USA). In addition, the results given here are based on experiments with a static economic model that is unable to simulate the effects of adjustments in technology and management that help USA agriculture adapt to changes of climate and thus mitigate some of their negative impacts. These adjustments are considered in Chapter 8.

Canada

Probably the most comprehensive of regional impact assessments, particularly in its attempt to simulate potential impacts on the entire provincial economy, is the IIASA/UNEP study in Saskatchewan.[8] It considered the implications of warming for future risk of soil erosion, for crop yields, for farm income (on farms of different size and on different soil types throughout the province), for gross domestic product of the province and for total provincial employment. This involved linking crop-response models with farm-level economic models so that information on altered yields could be used as inputs to an analysis of farm profitability and farm output. Information on farm output for each type of farm size and soil type, but aggregated over the entire province, was then fed into an input–output model that simulated how non-agricultual sectors might respond when agricultural-activity rates are altered (for example, as a result of changes in the amount of fertilizer purchased by farmers, or in the amount of grain to be stored, insured and ultimately shipped).

Like the economic model used in the US study, the input–output model used in Saskatchewan was also unable to consider adaptation over the medium or long term. The effect of changes in one sector on another was represented by a statistical relationship established from analysis of data for the 1970s and 1980s, and it is certain that this would not hold for the future. The estimates of impacts on employment and the GNP are therefore subject to considerable uncertainty, and should be taken simply as being indicative of the direction and order of magnitude of possible effects, rather than as a detailed projection of them.

Figure 6.2 summarizes estimated effects on agriculture in Saskatchewan under a range of different climatic scenarios. As with the other IIASA/UNEP case studies, a comparison was made of effects under the equilibrium GISS 2 × CO_2 climate with effects of the extreme weather typical of individual years and periods in the recent past. In the Saskatchewan study comparison was made with the unusually dry year of 1961 and dry period 1933–37. Because the GISS

Figure 6.2 Estimated effects of climatic changes on agricultural production in Saskatchewan. Baseline climate is 1951–80 unless otherwise stated. (*Source:* Parry and Carter, 1988).[9]

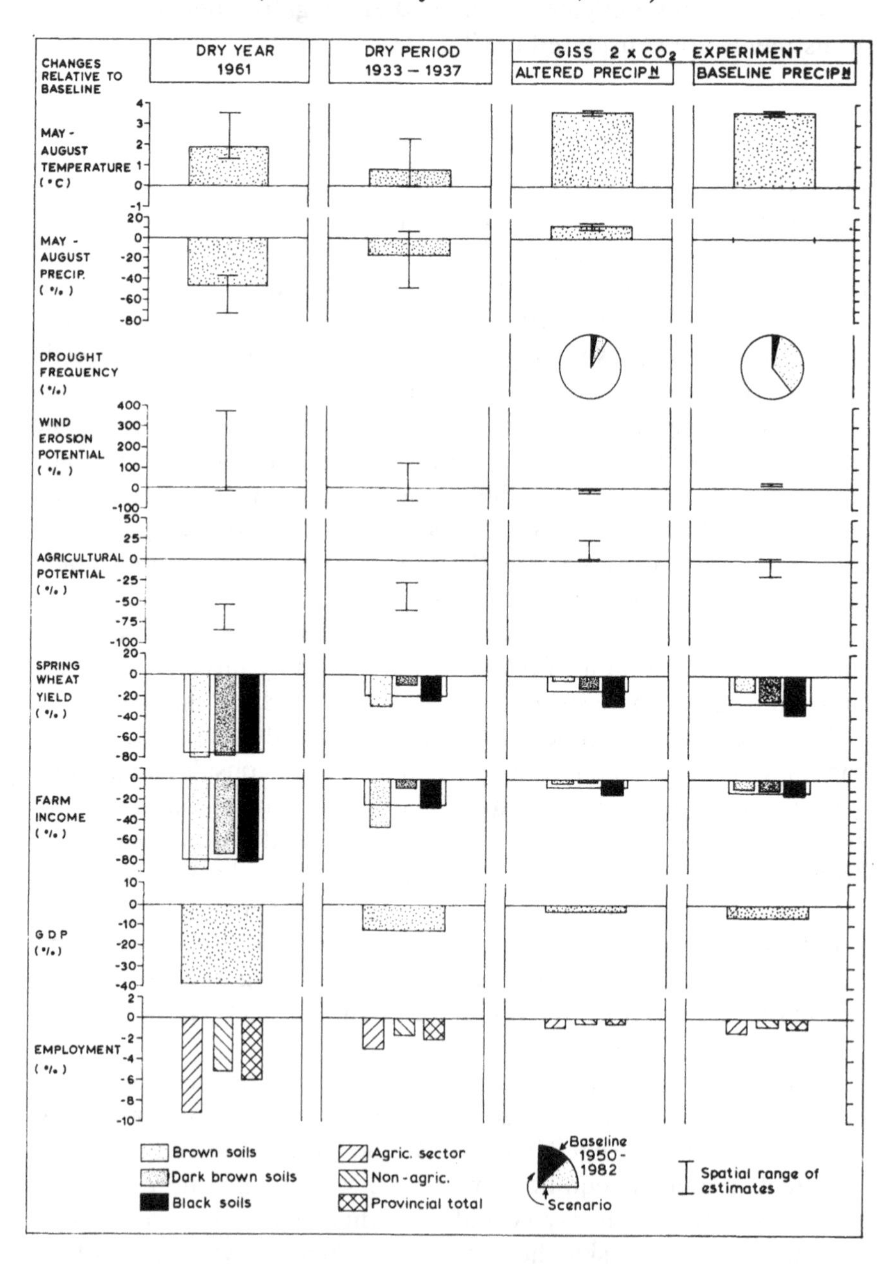

GCM may overestimate the increase in precipitation under a $2 \times CO_2$ climate, effects were estimated both for the modelled precipitation increases and for increases in temperature with no corresponding increase in precipitation.

With a 3.5 °C increase in growing-season temperatures but assuming no change in precipitation, wind-erosion potential increases by about a quarter, and the frequency of drought increases 13-fold. Spring-wheat yields are reduced by 15–37 per cent depending on the type of soil, and overall provincial output falls by 28 per cent. Since Saskatchewan at present produces 18 per cent of all the world's traded wheat, such a reduction could well have global implications (see Chapter 7). Calculations such as these assume that the present-day allocation of land to spring wheat is maintained in the future, and that today's cultivars are grown in the same way as at present. These are clearly unrealistic assumptions which need to be reconsidered in the light of possible technological and economic adjustments to climatic change (see Chapter 8).

Assuming (unrealistically) that the present-day relationship between production and profit holds in the future, average farm-household income is estimated to fall by 12 per cent, resulting in a reduction in expenditure by agriculture of Can.$277 million on the goods and services provided by other sectors, leading to a Can.$250 million (6 per cent) reduction in provincial GDP in sectors other than agriculture and a 1 per cent loss of jobs.

In general it seems that an average weather-year under the GISS $2 \times CO_2$ climate would be broadly similar in its scale of effect on wheat yields and wheat production to that of the most extreme dry period on record, 1933–37. Individual extreme years, such as 1961, have exceeded this level of impact. It is probable, however, that a broadly similar range of extremes would characterize a warmer and drier climate, so that anomalies such as the dry year 1961 would be likely to occur more frequently and quite possibly to be surpassed by even drier years than have recently been experienced in this region.

In Ontario the GFDL $2 \times CO_2$ climate implies increases of mean temperature during the growing season of about 1.7°C and precipitation in the growing season of about +45 per cent; and under the GISS $2 \times CO_2$ climate 1.8°C and 57 per cent. Under both these scenarios precipitation increases are more than offset by increases in evapotranspiration with consequent increased moisture stress on crops. Maize and soybean thus become very risky in the southern part of the province. In the north, where maize and soybean cannot currently be grown commercially because of inadequate warmth, cultivation may become profitable, but this is not expected to compensate for reduced potential further south and, if there were no adjustment of current land use and farming systems, the overall cost in lost production is reckoned at Can.$170 million and Can.$101 million under the GFDL and GISS 2 × CO_2 climates respectively.[10] However, warming could open up opportunities in southern Ontario for more extensive fruit and vegetable production which, with adequate new irrigation, could not only substitute for present grain farming but perhaps increase farm income. Much would depend, however, on the cost of irrigation at a time when other sectors might well also be drawing upon a diminished water supply.

Japan

A broadly similar hierarchy of models was used in an assessment of impacts on agriculture in Japan.[11] Here, experiments with a variety of rice models gave an indication of yield responses to climatic warming, and the altered yield levels were used as inputs to a national economic rice model that simulated how national production, prices and stocks of rice would respond to changes in yield, world prices and demand. Since the model assumes, unrealistically, that relationships between these factors will remain roughly similar in the future, the results of this study must be taken only as an indication of the direction and approximate magnitude of the potential impacts rather than as a detailed projection of them.

As in the Saskatchewan study, impacts under a GISS 2 × CO_2 climate were compared with those estimated for the weather of recent anomalous years or periods, such as the exceptionally warm year 1978 and cool year 1980, in order to gauge whether the scale of possible 2 × CO_2 climate impacts lies within or outside the range of recent (though only short-term) experience.

The effects on rice yields of higher temperatures assumed under the GISS 2 × CO_2 scenario are of a similar magnitude to those estimated for 1978 (Figure 6.3). In central Japan (Tohoku) rice yields increase about 8 per cent in an exceptionally warm year such as 1978, and about 2 per cent under the 2 × CO_2 climate. As a consequence, assuming that the rice area remains unaltered, the district rice supply is estimated to increase by 10 per cent and 7 per cent, respectively.

It should be noted that we are also assuming, in these estimates, that current rice cultivation techniques, including the use of

Figure 6.3 Estimated effects of climatic variations on agricultural production in Hokkaido (Hokk.), Tohoku (Toho.) and all Japan. Baseline climate is 1951–80, unless otherwise indicated. (*Source:* Parry and Carter, 1988).[9]

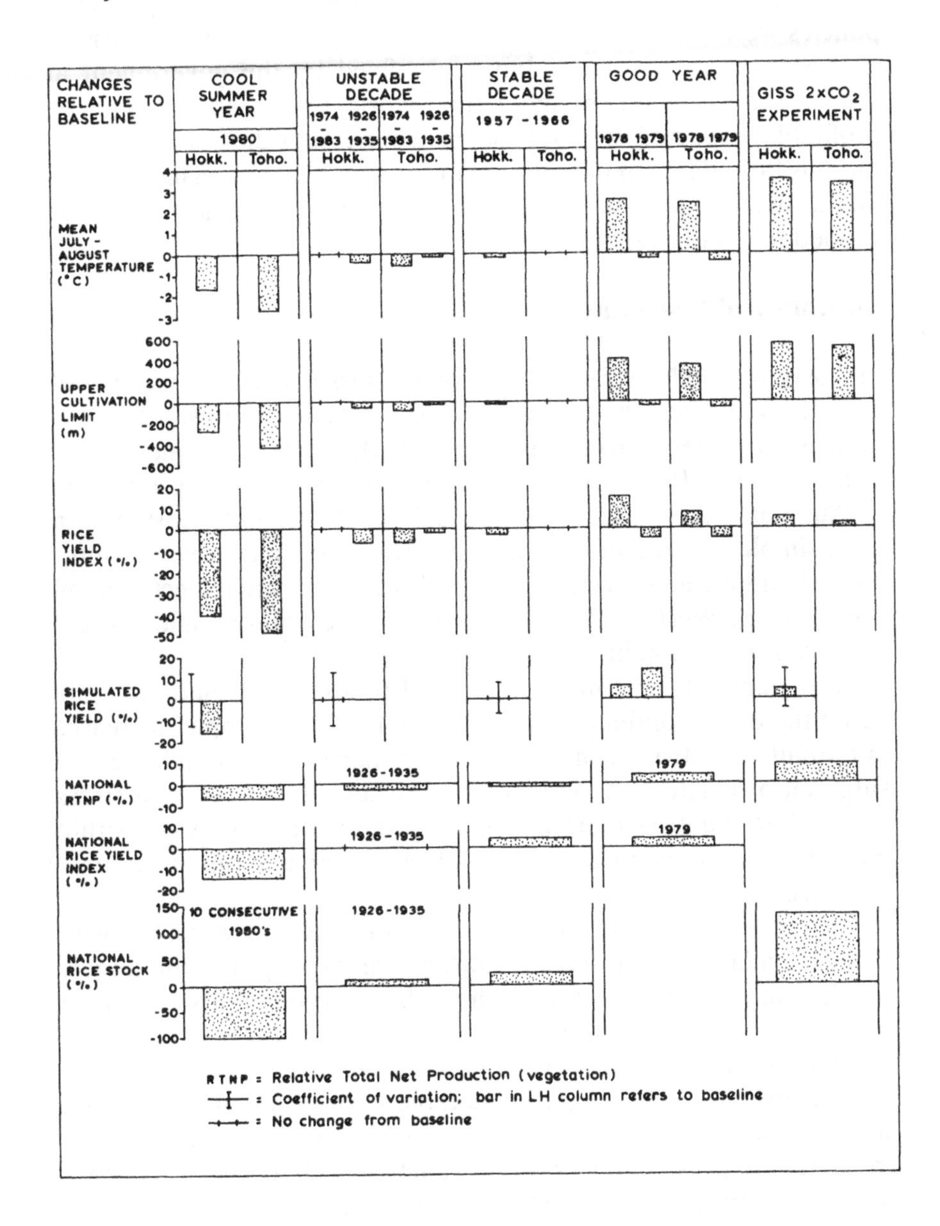

early-maturing rice cultivars, are retained in use under the warmer climatic conditions, which is clearly unrealistic since farmers will almost certainly adapt their husbandry to the altered conditions. But even if they retained their present, and inappropriate, cultivars Japan would produce 1–2 million tonnes more of rice each year and national rice stocks would double over a 16-year period if policies were not introduced to discourage production. At present, domestic prices of rice in Japan are fixed by the government at a level some five times higher than world prices, so it cannot be profitably exported without heavy export subsidies. In addition, there is little demand outside Japan for Japonica rice. This suggests that a substantial reduction in guaranteed domestic prices would be needed to avoid oversupply.

Australia and New Zealand

In Australia and New Zealand national assessments have been based on a thorough review of existing knowledge and on use of expert judgement rather than on experiments with a suite of agronomic and economic models.[4, 5] Overall, it is reckoned that wheat production in Australia could increase under a 2 × CO_2 climate, assuming a quite simple scenario of increased summer rainfall, decreased winter rainfall and a general warming of 3°C. Increases are expected in all states except Western Australia, where more aridity might cause a significant reduction in output.[12]

More generally, the major impact of production would probably be on the drier frontiers of arable cropping. For example, increases in rainfall in subtropical northern Australia could result in increased sorghum production at the expense of wheat. Increased heat stress might shift sheep farming and wool production southward within Australia, with sheep possibly replacing arable farming in some southern regions.

The cattle industry is probably more adaptable to warming, with breeds available to suit most conditions, and changes would largely be determined by prices and the relative profitability of wheat farming or wool production.

Many areas currently under production for apples, pears and stone fruits would no longer be suitable under a 3°C warming, and would need to shift southwards or to higher elevations in order to maintain present levels of production. All of these changes would also be affected by changes in the distribution of diseases and pests.

Much would depend, however, on how the El Niño Southern Oscillation (ENSO) system responds to global warming. El Niño, named after the Christ-child because of its tendency to occur around the end of the year, is a temporary warming of the eastern tropical Pacific Ocean associated with changes in atmospheric circulation in the western Pacific. It dominates year-to-year variability in rainfall over most of northern and eastern Australia. When there is a major ENSO event, as in 1982/83, there are serious droughts in these regions. At present we have no clear picture as to whether these events will occur more or less frequently as the climate warms.

In New Zealand, with temperature increases of 3°C to 5°C estimated for a doubling of GHG, production of wheat would probably be reduced in North Island and in the centre of South Island, but would be increased at higher latitudes in the Otago-Southland region.[5] Production of maize would be likely to increase as it extends its range southwards, but traditional production areas in North Island could come under competitive pressure from higher value horticultural crops, forcing maize onto land with lower productive-potential and thus possibly decreasing average yields. Higher temperatures would lead to increased upland pasture production throughout New Zealand, increased winter-pasture production in the lowlands but possibly decreased summer production.

EFFECTS ON PRODUCTION IN NORTHERN MARGINAL REGIONS

In addition to studies in Canada and Japan, the IIASA/UNEP project also conducted surveys on likely impacts in three northern high-latitude regions – Iceland, Finland and the north European region of the USSR. Their aim was to evaluate the effect of warming in these currently cold regions, and they pursued a common strategy that involved the use of hierarchies of models (of crop response, farm response, and regional impact) and compared potential impacts under the GISS 2 × CO_2 climate with impacts estimated under the weather of recent extremely warm and cold years and periods in order to gauge the significance of the estimated 2 × CO_2 impacts.

Iceland

In Iceland, a linked series of hay-yield and livestock models sought to calculate the effect of warming on the number of sheep that the island could carry on its pastures (sheep farming accounts for

Figure 6.4 Estimated effects of climatic variations on agricultural production in Iceland. Baseline climate is 1951–80. H = high input (120 kg/ha nitrogen); L = low input (80 kg/ha); Ca = cattle; Ho = horses; Sh = sheep; 1 = from national model; 2 = from refined national model. (*Source:* Parry and Carter, 1988).[9]

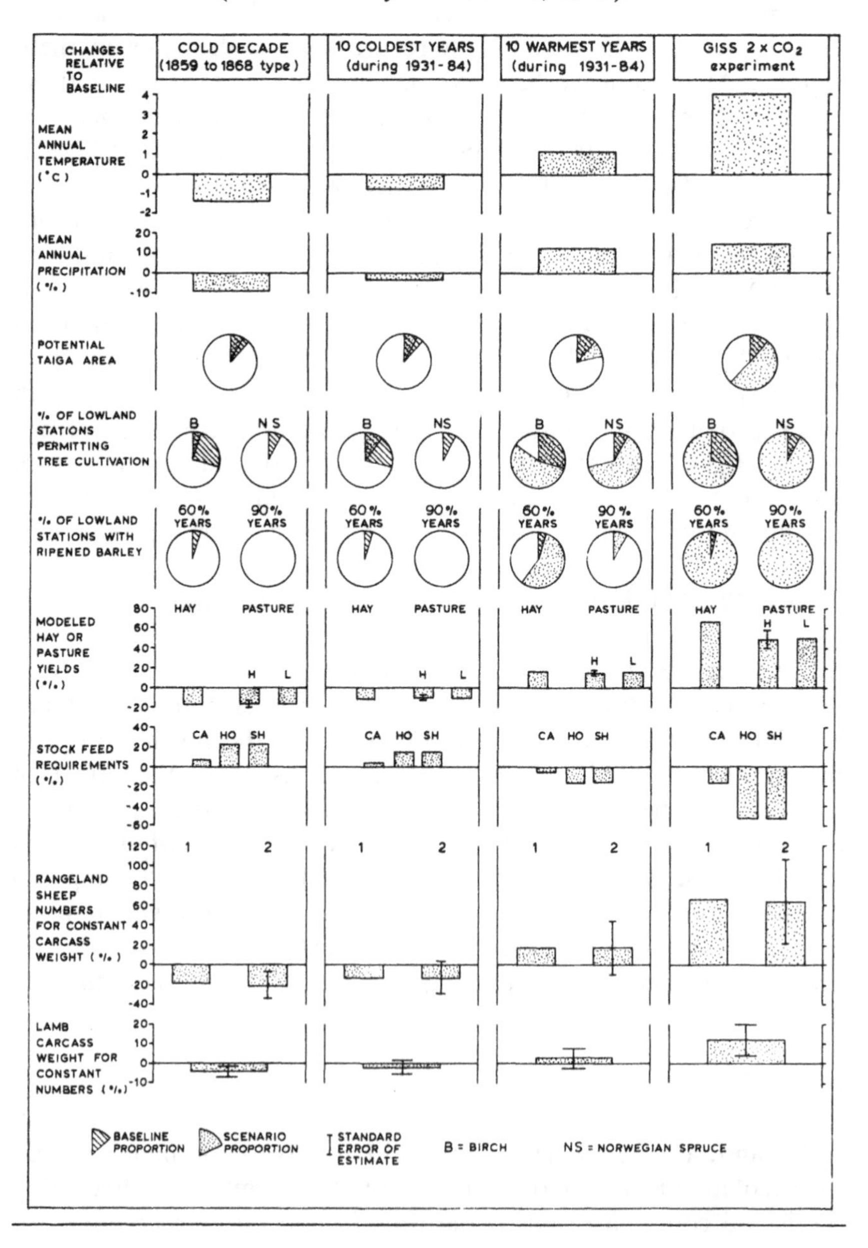

about three-quarters of Icelandic agriculture by value).[13] With mean annual temperatures increased by 4.0°C and precipitation 15 per cent above the present average (consistent with the GISS 2 × CO_2 climate), the onset of the growing season of grass is brought forward by almost 50 days, hay yields on improved pastures increase by about two-thirds and herbage on unimproved rangelands increases by about a half (Figure 6.4). The number of sheep that could be carried on the pastures is raised by about 250 per cent and on the rangelands by two-thirds if the average carcass weight of sheep and lambs is maintained as at present. If, however, sheep numbers were kept at their present number then the average carcass weight of lambs could be expected to increase by over ten per cent. The magnitude of these impacts is about four-times that of impacts estimated under the weather of the ten warmest years in recent history.

At a guess, output of Icelandic agriculture could probably double with a warming of 4°C. At the same time farmers could expect to make substantial savings (over 50 per cent) on the amount of bought-in feed needed for overwintering flocks because of the longer grazing season. At present 1–2 per cent of Iceland's GNP is spent on the import of fodder.[13]

Savings of about 10–15 per cent could also probably be made on the reduced import of timber requirements (currently 2 per cent of all imports) by afforestation, because warming is estimated to enable the whole of Iceland to be within the potential growing area for spruce under the GISS 2 × CO_2 climate in contrast to only 8 per cent under the present climate.

However, a warmer climate could well increase the danger of diseases in plants (e.g. potato blight) and animals (e.g. parasitism), thus affecting fodder-crop yield and the volume of livestock production.

Finland

In Finland a combination of crop-yield and farm-income models was used to estimate the possible effects of climatic changes on farm profitability.[14] Assuming an increase in summer warmth by about one-third and precipitation by about half (consistent with the GISS 2 × CO_2 climate), barley and spring-wheat yields increase by about 10 per cent in the south of the country but by slightly more in the north (due to relatively greater warming and lower present-day yields). If we also assume (unrealistically) that the current relationship

Figure 6.5 Estimated effects of climatic variations on agricultural production in northern (N), central (C) and southern (S) Finland. Baseline climate is 1959–83 or 1971–80, as shown. 1 = estimated farm income based on yield estimates from all regions. (*Source:* Parry and Carter, 1988).[9]

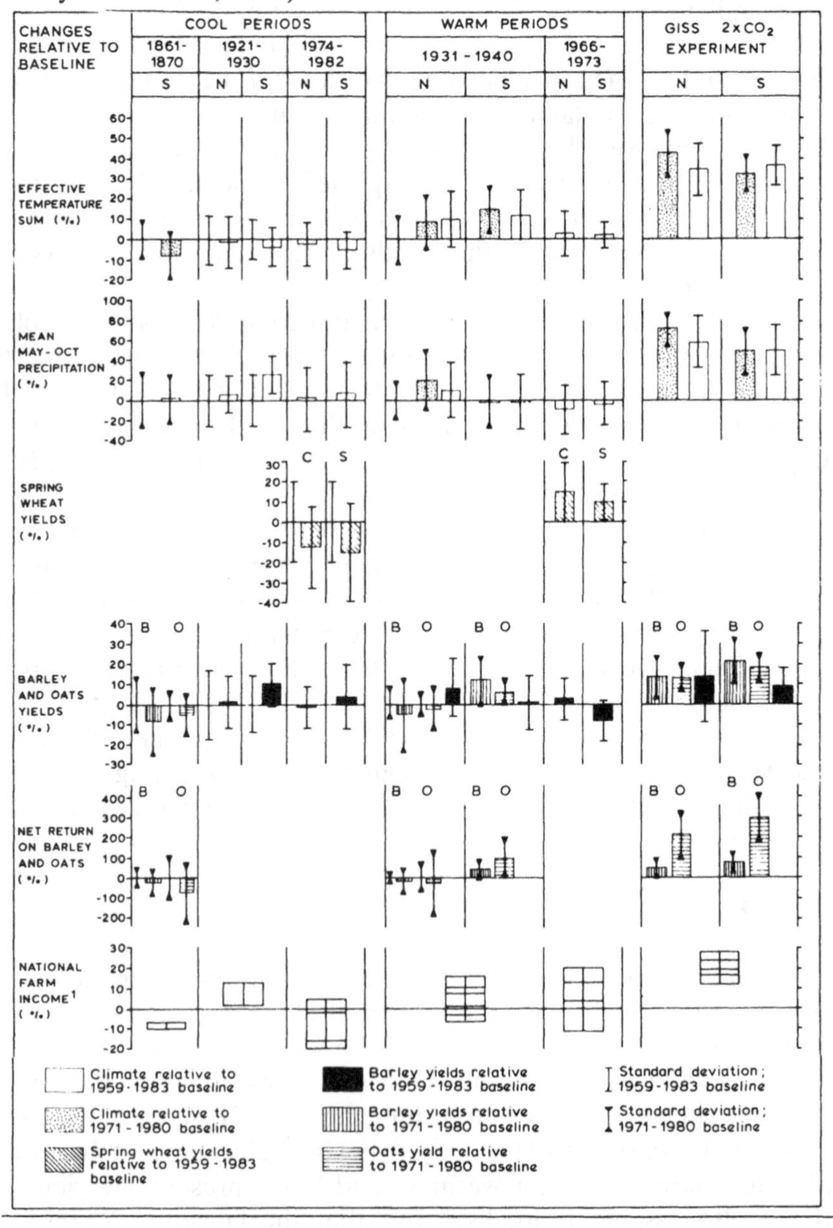

between yield, production and profitability holds in the future as it does now, then the net return on barley yields will increase by about three-quarters in the south, and national farm income will be raised between 10 and 25 per cent (Figure 6.5). These impacts are about double those estimated to occur in the weather typical of the warmest periods that have occurred this century (e.g. 1931–40 and 1966–73).

The area under grain production in Finland might increase at the expense of grass and livestock production as a consequence of raised profitability, with the greatest extension being in winter crops such as wheat rather than spring crops such as barley or oats. The growing area of wheat, rye, oilseeds and sugarbeet, which at present compete for limited land in the south, would extend further northwards, and Finland could reasonably be expected to become self-sufficient in bread grains, which it has not quite been in recent years.

Greater increases in yields in the north would assist present current regional policy which aims to reduce differences in farm incomes between north and south. At present, prices of farm inputs are subsidized in the north and, assuming that this regional policy is not altered, northern farmers would stand to benefit most. The probability is that regional policies would be adjusted to reflect the altered balance of profitability between north and south, and to avoid excess production.

Northern USSR

The only other region for which an integrated impact assessment has been completed is in the north European USSR.[15] Here Soviet scientists on the IIASA/UNEP project investigated potential effects on agriculture in three areas near the current northern limit of arable farming: in the region around Leningrad (60°N), in the so-called Central Region (around Moscow, 55°N), and in Cherdyn which lies on the western side of the Ural mountains (60°N). A variety of GCM-based and synthetic climatic scenarios were adopted, including the GISS 2 × CO_2 equilibrium climate, a simple linear transient response of climate that assumed an equilibrium 2 × CO_2 condition in 2050, and climates that were systematically +1°C and +1.5°C warmer than the present.

Results of the USSR case studies are summarized in Figure 6.6. In the Leningrad and Cherdyn regions, under climates that are 2.2°C–2.7°C warmer during the growing season and 36–50 per cent wetter (consistent with the GISS 2 × CO_2 climate), yields

Figure 6.6 Estimated effects of climatic variations on agricultural production in the northern European USSR. Baseline climate is 1951–80 for the Leningrad and Cherdyn regions and 1931–60 for the Central Region. 1 = climatic change estimated using an empirical method (Vinnikov and Groisman, 1979)[16]; 2 = assumed date of CO_2-doubling; 3 = relative to technology trend. (*Source:* Parry and Carter, 1988).[9]

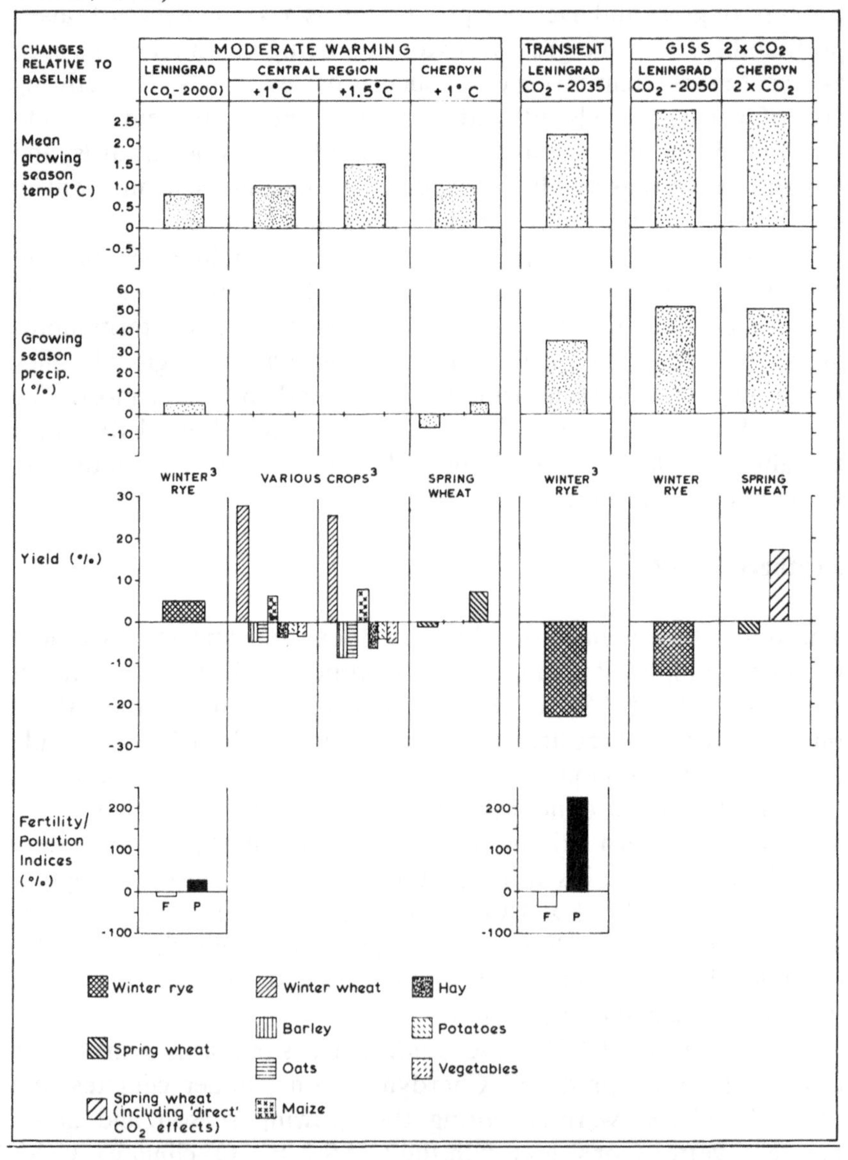

vary greatly according to crop type. For example, in the Leningrad region winter-rye yields are estimated to decrease by about a quarter due to faster growth and increased heat stress under the higher temperatures.

However, crops such as winter wheat and maize, which are currently low-yielding because of the relatively short growing season in these regions, are better able to exploit the higher temperatures and exhibit yield increases in Cherdyn of up to 28 per cent and 6 per cent respectively, with a 1°C warming.

The differential yield responses described above are reflected in substantial changes in production costs incurred in meeting production targets. Thus, while production costs for winter wheat and maize in the Central Region around Moscow are reduced by 22 per cent and 6 per cent under a 1°C warming and with no change in precipitation, they increase for most other crops, particularly quick-maturing spring-sown ones which are the dominant crops today. This would suggest that quite major switches of land use would result and the land allocation models used in the study indicate that, to optimize land use by minimizing production costs, winter wheat and maize would extend their area by 29 per cent and 5 per cent, while barley, oats and potatoes would decrease in extent. Such changes in land use as a response to change in climate are considered in more detail in Chapter 8.

Overall, conditions under the GISS 2 × CO_2 climate are detrimental. They are either too wet, leading to nutrient depletion and waterlogging, or too warm, leading to premature crop-ripening. Production costs increase for the major current crops and it is probable that total output from current farming systems in northern regions of Soviet arable agriculture would decrease. A substantial northward shift of farming types, probably of about 500 km, would be necessary to accommodate the changes in temperature and rainfall projected for a doubling of GHG, thus taking advantage of the increased thermal resources for agriculture at higher latitude. This might well be needed to offset the losses in production threatened by higher temperatures and increasing aridity that could occur in the current winter-wheat and maize growing regions further south. Whether soils and terrain would permit such a shift has not yet been investigated.

CONCLUSIONS

The effects of possible climatic changes on regional and national production in agriculture have not yet been investigated in any great detail, nor for more than a few case studies. No comprehensive studies have been made in any tropical region.

The effects are strongly dependent on the many adjustments in agricultural technology and management that undoubtedly will occur in response to any climatic change. So numerous and varied are these potential adjustments that it is extraordinarily difficult to evaluate their ultimate effect on aggregate production. In this chapter we have therefore considered the effects on production that are likely to stem directly from changes in yield, unmodified by altered technology and management. Adjustments in technology will be considered in Chapter 8.

In summary, it seems that overall output from the major present-day grain-producing regions could well decrease under the warming and possible drying expected in these regions. In the USA, grain production may be reduced by 10–20 per cent and, while production would still be sufficient for domestic needs, the amount for export would probably decline. Production may also decrease in the Canadian prairies and in the southern USSR. In Europe production of grain might increase in the UK and the Low Countries if rainfall increases sufficiently, but may be reduced in southern Europe substantially if there are significant decreases in rainfall, as currently estimated in most GCM experiments. Output could increase in Australia if there is a sufficient increase in summer rainfall to compensate for higher temperatures.

Production in semi-arid and humid tropical regions is most affected by precipitation and we have little confidence in current estimates as to how this may change.

Production could increase in regions currently near the low-temperature limit of grain growing, in the northern hemisphere in the northern Canadian prairies, Scandinavia, north European USSR, and in the southern hemisphere in southern New Zealand, and southern parts of Argentina and Chile. But it is reasonably clear that, because of the limited area unconstrained by inappropriate soils and terrain, increased high-latitude output will probably not compensate for reduced output at mid-latitudes. The implications of this for global food supply and food security are considered in the next chapter.

7. IMPLICATIONS FOR GLOBAL FOOD SECURITY

Although, on average, global food supply currently exceeds demand by about 20 per cent, its year-to-year variation (which is about + or − 10 per cent) can reduce supply in certain years to levels where it is barely sufficient to meet requirements. In addition, there are major regional variations in the balance between supply and demand, with perhaps a billion people (about 15 per cent of the world's population) not having secure access to sufficient quantity or quality of food to lead fully productive lives. For this reason the working group on food security at the 1988 Toronto Conference on The Changing Atmosphere concluded that: "While averaged *global* food supplies may not be seriously threatened, *unless appropriate action is taken to anticipate climate change and adapt to it*, serious regional and year-to-year food shortages may result, with particular impact on vulnerable groups".[1] Statements such as this are, however, based more on intuition than on knowledge derived from specific study of the possible impact of climatic change on food supply. No such study has yet been completed, although one is currently being implemented and the report is due in 1992.[2]

The information available at present is extremely limited. It has for example, been estimated that increased costs of food production due to climatic change could reduce *per capita* global GNP by a few percentage points.[3] Others have argued that technological changes in agriculture will override any negative effects of climatic changes and, at the global level, there is no compelling evidence that food supplies will be radicaly diminished.[4] Recent reviews, including that by the IPCC, have tended to conclude however that, at a regional level, food security could be seriously threatened by climatic change, particularly in less developed countries in the semi-arid and humid tropics.[5, 6] An important next step is to conduct a sytematic and careful analysis of the sensitivity of the world food trade to climatic change.

MODELLING THE IMPACT OF CLIMATE CHANGE ON THE WORLD FOOD SYSTEM

There are a variety of models that attempt to simulate the mechanisms of world trade in agricultural commodities. These vary from systems-dynamic methods to quite static input–output formulations. Their time horizons vary from 10 to 200 years, and their geographical aggregation ranges from a single world unit to over 100 different countries.

In most cases the potential effects of an altered climate can be analysed by exogenously manipulating yield components in the models, but there are two major obstacles to this. First, the models were not designed to simulate the effects of climatic change and may not respond sensibly to large-scale differential changes in yield. Secondly, the more comprehensive of the models require estimates of yield changes for a large number of countries and for all agricultural commodities, and this level of detail of yield responses to projected changes of climate is not currently available.

It is largely as a consequence of these difficulties that very few explicit climatic "experiments" have been conducted with global food models. Two preliminary studies have been published: one conducted at the US National Center for Atmospheric Research (NCAR) and the other at the US National Defense University (NDU).[8, 9, 10]

The NCAR study used the International Futures Simulation Model (IFS) in which climate is represented by a yield factor. The yield factor was varied as a surrogate for climatic change to examine the response of production, exports, imports, crop prices, reserve levels, global starvation, etc. These model runs with altered yield were then compared with runs assuming no yield changes due to climatic change.

Two "perturbed" runs were made: one with the model's "climate" (the yield factor) being gradually altered beginning in 1985 and reaching the maximum alteration of 20 per cent in the year 2000. The model predicted changes in global crop production of 5–7 per cent in both directions by the year 2000, in effect estimating that the world agricultural system had the capacity to absorb about two-thirds of the potential of a slow change in climate by adjusting land area under production, land area under different crops and the intensiveness of production as a response to altered crop prices.

But less stability was evident in the face of an immediate, short-term change in yield, such as could result from a pattern of adverse weather events (e.g. concurrent droughts in the major mid-latitude grain-producing regions). A −20 per cent yield in one year created a major perturbation in production, with world crop reserves reduced almost to zero. Overcompensating increases in production in the following year created glut and price collapse (Figure 7.1).

An important conclusion emerging from these results is that year-to-year variations in yield due to weather could be an important

Figure 7.1 Simulated agricultural effects of perturbed "climate" versus control runs to the year 2000 using the International Futures Simulation Model. (a) crop yields in the USA with slow trend change in yield factor to −20%; (b) world crop production with slow trend changes in yield factor to ±20%; effect on world crop production (c) and reserves (d) of a single −20% pulse in 1985. (Adapted from Warrick *et al.*, 1986).[7]

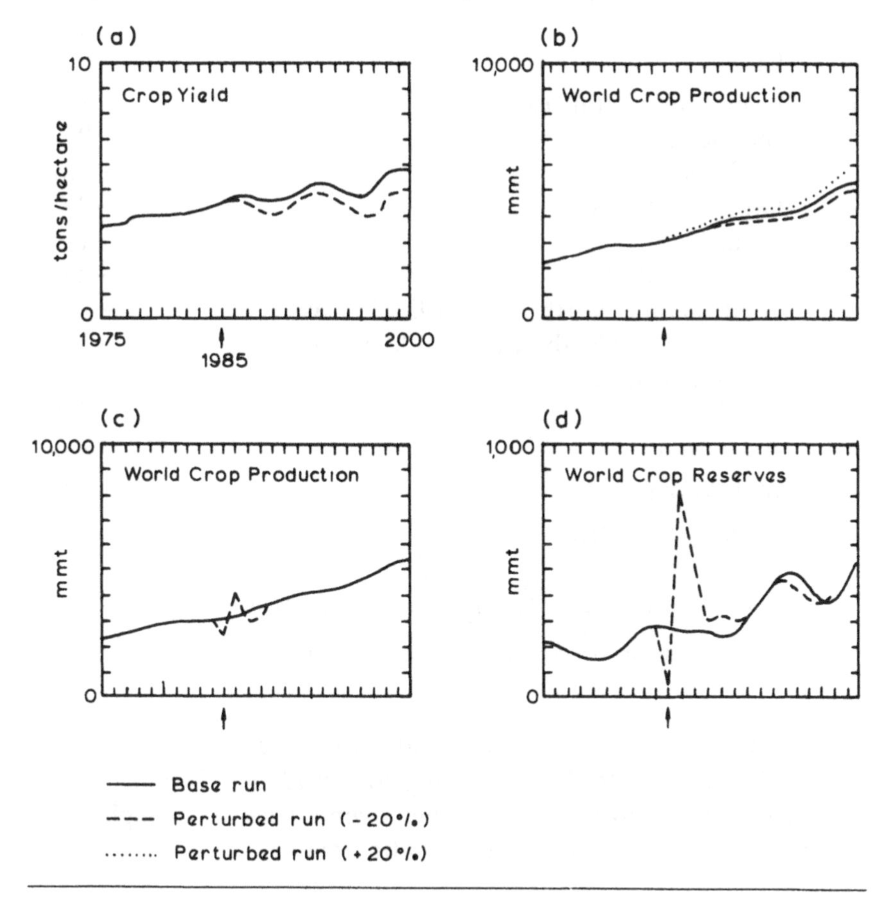

Table 7.1: Simulated global grain production in the year 2000 under a large warming scenario, as a percent of base level projections[1]

Group/Country	*% from base level*
I. Developed Countries:	−2.4
United States	−3.8
Canada	6.0
European Community	−2.2
Other Western Europe	−2.0
Australia	−3.0
South Africa	−4.1
II. Centrally Planned Countries:	3.1
Eastern Europe	1.1
USSR	6.1
China	0.7
III. Developing Countries	−1.4
Indonesia	0.5
Thailand	−3.6
Other South-east Asia	−0.0
India	−1.8
Other South Asia	−2.0
High Income North Africa/Middle East	−2.8
Low Income North Africa/Middle East	−2.9
Central America	−2.8
Brazil	0.3
Argentina	2.6
IV. Total Above	0.0
V. Warming Countries, Total[2]	3.3

Source: Warrick *et al.*, 1986,[7] adapted from NDU, 1983[10]

[1] Large warming scenario (ΔT, ΔP) = 1.4°C, 6% high-mid latitudes: 1.0°C, 2% mid-low latitudes: 0.75°C, 2% subtropics.

[2] Countries favourably impacted by warming (Canada, Eastern Europe, USSR, China).

source of instability in world food supply, even if changes in mean climate due to GHG forcing are gradual. Changes in the frequency of yield-reducing weather events such as droughts and warm spells due to long-term climatic change, about which very little is at present known, are thus likely to be critical to food security.

The study by the National Defense University considered a range of climatic scenarios from large cooling to large warming and derived estimates of yield changes, region by region, for each scenario on the basis of expert opinion. These were then used as inputs to an economic equilibrium model of the agricultural sector (the USDA Grains-Oilseeds-Livestock or GOL model).[10] The model was used to estimate changes in grain production that resulted from changing patterns of comparative advantage, changing investment levels and shifts in land use as the model attempted to balance supply and demand. The production estimates for a large warming scenario are given in Table 7.1. Under this scenario some countries such as the USSR and Canada gain appreciably while others have reduced production (e.g. USA, Australia, South Asia). Overall, the experiment suggests no change in net global grain-production.

These results should not, however, be taken as an indication of what might happen under a climate altered by greenhouse gases. The climatic scenario adopted in Table 7.1 is both highly generalized and assumes a relatively small amount of warming. It does not allow for regional decreases in precipitation that may occur particularly in some mid-continental, mid-latitude areas. Moreover, the estimates of yield responses to altered climate are not derived from experiment or simulation but from expert judgement.

A more broadly based analysis of impacts on the global agricultural sector was conducted for the Canadian Atmospheric Environment Service (AES) in order to assess the implications of warming for the competitive position of Canadian agriculture.[11] This was based on a survey of all available literature on the likely regional responses of agriculture to climatic change (such as the IIASA/UNEP study), interpreted in the context of a global climate warming scenario that specified both a range of temperature changes and also changes in available moisture, based upon the outputs of experiments with general circulation models (Table 7.2).

Table 7.3 gives the changes in production opportunities for each crop in each region estimated by the Canadian study. These are based on a survey of recent impact studies and represent effects on crops that are currently grown. The table does not assume a change in crops or production technologies, nor does it account for the effects of sea-level rise, or secondary impacts stemming from effects of climatic changes on pests and diseases.

Table 7.2: Global Climate Warming Scenario used by Canadian AES Study

Region	*Relative temperature increases*	*Relative change in moisture*
Canada		
North	Large	Wetter (?)
South	Large	Drier
Rest of North and Central America		
United States	Moderate	Drier
Mexico	Small	Wetter
South America		
North	Small	Drier in most regions
South	Moderate	except central region which may become drier
Europe		
North	Large	Wetter
South	Moderate	Drier (?)
Africa	Small	Wetter in most regions except central region which may become drier
USSR		
North	Large	Wetter (?)
South	Large	Drier
Asia		
China	Moderate	Drier
India	Small	Wetter
Oceania	Small	Wetter

Source :Smit, 1989.[11]

Production opportunities for wheat are estimated to decrease in most regions due to an increase in temperatures beyond levels optimal for wheat growth. Increases estimated for Canada and the USSR are based on results of recent impact studies that reveal a high degree of sensivity of wheat yield to changes in precipitation.

Table 7.3: Changes in production opportunities estimated by Canadian AES Study

Region	*Crops*					
	Wheat	*Grain Corn*	*Barley*	*Oats*	*Soybeans*	*Rice*
Canada	▲	▲	▼	▼	▼	na[1]
Rest of North and Central America	▼	▼	na	na	▼	na
South America	▼	▼	na	na	▼	na
Europe	▼	▼	▼	▼	na	na
Africa	▼	na	na	na	na	na
USSR	▲	▲	▼	▼	na	na
Asia	▼	na	na	na	na	▲
Oceania	▼	na	na	na	na	na

[1] Not available

Note: These estimates of change (▲ represents an increase, ▼ represents a decrease) in production opportunities associated with climate warming are based upon interpretation and synthesis of independent climate impact studies, and are supplemented by expert opinion.
Source: Smit, 1989.[11]

The positive impacts reflect precipitation increases assumed in the studies' climatic scenarios and, since these projections for precipitation are most uncertain, any conclusion concerning changes in production opportunities are speculative. Similarly uncertain, and for the same reason, are the estimates of increased production opportunities for maize in Canada and the USSR.

SENSITIVITY OF THE FOOD SYSTEM

A major drawback of the studies considered above is their focus on a single scenario of climatic change and a single response of yield to that change. In reality there is a wide range of uncertainty concerning both the climate response to GHG forcing and the yield response to a consequent change of climate. Given this uncertainty

Table 7.4: Assumed changes in yield (%) under an altered climate (for explanation, see text)

Scenario I : Estimate of moderate impacts

	Wheat	*Maize*	*Soybeans*	*Rice*	*Other crops*
USA	−10	−15	−15	0	−10
Canada	−15	+5	+5	0	−10
EC	−10	0	0	0	−5
Australia	+10	+10	+10	+15	+10
China	+10	+10	+10	+10	+10
USSR	+10	+15	+15	0	+10
N. Europe	+15	+30	0	0	+10
Japan	−5	0	+15	+10	+5
Rest of World	0	0	0	0	0

Scenario II : Estimate of adverse impacts

	Wheat	*Maize*	*Soybeans*	*Rice*	*Other crops*
USA	−15	−30	−30	−10	−20
Canada	−20	0	0	0	−20
EC	−10	−10	−10	0	−10
Australia	−5	0	0	0	0
China	−10	−10	−10	−10	−10
USSR	−10	−5	−5	−5	−10
N. Europe	+10	+20	0	0	+10
Japan	−5	0	0	0	0
Rest of World	0	0	0	0	0

Scenario III : Estimate of very adverse impacts

	Wheat	*Maize*	*Soybeans*	*Rice*	*Other crops*
USA	−20	−40	−40	−15	−20
Canada	−20	−5	−5	0	−20
EC	−15	−10	−10	0	−10
Australia	−15	−10	−10	0	−15
China	−15	−15	−15	−15	−15
USSR	−15	−10	−10	−20	−15
N. Europe	+10	+15	0	0	+10
Japan	−5	0	0	−5	0
Rest of World	−10	−10	−10	−10	−10

it is preferable to estimate the sensitivity of the world food system to a range of possible changes in production potential.

Table 7.4 summarizes, region-by-region, estimates of changes in yield presented in the preceding chapters of this book. The estimates are for three magnitudes of possible impact (scenarios I,II and III). Scenario I represents an estimate of moderate impacts under a 2 × CO_2 climate, while scenarios II and III represent more adverse but still realistic impacts (for example those that might occur if precipitation increases in major producing regions were insufficient to compensate for increased rates of evapotranspiration and there consequently occurred significant reductions in crop water availability). The data presented here do not comprise forecasts of impacts, but are intended to represent the range of possible effects and thus provide a basis for assessing their effect on world food output and prices.

The estimates of regional production changes were used as inputs to an international trade model, the USDA Static World Policy Simulation (SWOPSIM). SWOPSIM is a partial-equilibrium, static model that provides a highly simplified representation of world agriculture.[12] Because it is not dynamic it can only provide a snapshot of the effects on world trade assuming that the yield changes occurred under present-day conditions of land use, technology, and trade arrangements. It cannot simulate how these conditions may themselves respond to changes in climate, in yields and in other factors.

Table 7.5 gives the changes in welfare estimated by the model, which reflect the simulated production and price responses to changes in yield. It should be emphasized that the numbers given here do not represent a forecast of impacts because they do not take into account the spontaneous adjustments that will occur within agriculture and will markedly affect how production responds to climatic change.

Under scenario I the price of maize and soybeans is estimated to increase by about 10 per cent, reflecting their location of production in mid-latitude regions that may be adversely affected by warming and drying. As a result, feed costs increase thus raising the price of most livestock products. Wheat and other coarse grain prices decrease slightly. Rice prices decrease on the assumption that the main rice-producing regions of the world would benefit from a limited amount of warming combined with moderate increases in rainfall.

The small price increases suggest that net food-production capability

Table 7.5: Change in welfare, by country or region, under three scenarios (I, II, III) of climatic change

	I	II	III
Countries/Regions	*Welfare as % of agricultural GDP*	*Welfare as % of agricultural GDP*	*Welfare as % of agricultural GDP*
United States	0.17%	7.3%	10.3%
Canada	1.18%	0.75%	5.25%
EEC	0.28%	3.5%	5%
Northern Europe	0.13%	0	1.25%
Japan	1.56%	2.5%	5%
Australia	0.54%	3.57%	0.57%
China	0.71%	1.35%	3.1%
USSR	1.95%	13.33%	43.2%
Brazil	0.15%	0	2%
Argentina	1.33%	15.67%	31.33%
Pakistan	0.57%	1.7%	6.78%
Thailand	0.32%	1.44%	4.88%
Rest of the World	0.01%	0.5%	4.2%
World total	0.1%	2.4%	4.7%

Source: Reilly, J. personal communication[13]

of the world changes little under scenario I, because positive production changes in some regions broadly compensate for negative production changes in other regions. For example, there occur increases in productive potential in Australia, China and the USSR that broadly match losses in North America and Europe. It should be recalled, however, that the basis for this assumption is weak and that regionally-compensating losses and gains of this kind may not necessarily occur. In addition, the changes in potential production are not the most negative that could occur. Under scenarios II and III total world loss of welfare measured as percentage change in agricultural GDP amounts to 2.4 per cent and 4.7 per cent of gross income, respectively, as compared with 0.1 per cent under scenario I (Table 7.5). Thus there is an order of magnitude difference in net impact between the losses of productive potential estimated for a moderate impact under a 2 × CO_2 climate (scenario I) and the larger losses that could occur if moisture increases did not match temperature increases.

These impacts may become especially severe if there are concurrent reductions in productive potential in the main grain-exporting regions (USA, EC and Canada). In each of these regions, for four scenarios of climatic change, the SWOPSIM model was run a number of times to provide an indication of the sensitivity of agricultural prices and welfare to progressively-decreasing yields that were assumed to reflect the possible effects of progressive mid-latitude, mid-continent drying.

The scenarios are given in Table 7.6. Scenario A assumes no yield changes in the rest of the world. The other scenarios assume various changes in another set of regions (USSR, China, northern Europe, Australia, Argentina, Japan and Brazil). Scenarios B and C reflect the possibility that agriculture in these areas may benefit from higher temperatures and increased precipitation.

Table 7.6: Scenarios of food output under altered climate (for explanation, see text)

Scenario	*Country/Region*	*Change in output*
A	USA, Canada, EC	−10% to −70%
	Rest of World	No change
B	USA, Canada, EC	−10% to −70%
	USSR, China, Northern Europe, Argentina, Brazil, Japan, Australia	+25%
	Rest of World	No change
C	USA, Canada, EC	−10% to −70%
	USSR, China, Northern Europe, Argentina, Brazil, Japan, Australia	+25%
	Rest of World	−25%
D	USA, Canada, EC	−10% to −70%
	USSR, China, Northern Europe, Argentina, Brazil, Japan, Australia	No change
	Rest of World	−25%

Scenarios C and D provide an indication of global agricultural effects if developing countries, suffering from capital constraints, experience difficulty responding to the modest changes of climate indicated for these regions.

Figure 7.2 Estimated changes in price of primary agricultural products to a range of yield reductions in the USA, the European Community and Canada. For description of the scenarios see Table 7.6. (*Source:* Reilly, personal communication, 1989)[13]

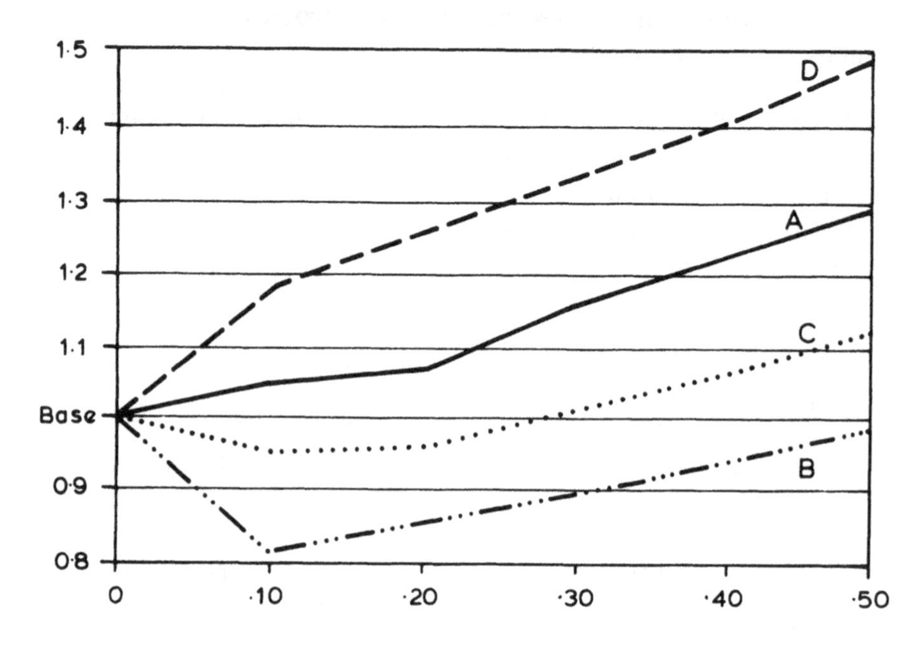

Yield Reduction in U.S.A., European Community and Canada

A: No other yield increases or decreases

B: Yield increases in Australia, U.S.S.R., China, South America and Northern Europe (+ 25 %)

C: Yield increases in Australia, U.S.S.R., China South America and Northern Europe (+ 25 %)
Yield decreases in Africa, South - East Asia (- 25 %)

D: Yield decreases in Africa, South - East Asia (- 25 %)

Aggregate crop-price effects are shown in Figure 7.2. As expected, prices increase as yields decrease in the mid-latitude grain-exporting regions. Only a small increase in prices occurs under scenario B, where enhanced yield effects in other regions partially offset the negative effects at mid-latitude. Indeed, there occurs an initial fall in aggregate prices due to decreases in the price of wheat and rice.

Under scenario A, (10–70 per cent yield reduction in the USA, the EC and Canada; no changes elsewhere) prices of primary products increase by 7 per cent above world levels with a 10 per cent yield reduction, and by 30 per cent with a 50 per cent yield reduction. Under the "worst case" scenario D (10–70 per cent yield reductions in USA, the EC, Canada; 25 per cent yield reduction in developing countries; no changes elsewhere) prices of primary products increase by 20 per cent and 50 per cent with a 10 per cent and 50 per cent yield reduction. The same trends are indicated for changes in world prices of meat, dairy products and oils, though the increases are smaller because prices are only indirectly affected through their dependence on crop production for feed.

Under scenario A, world economic welfare decreases by $5 billion (0.03 per cent of GDP) for a 10 per cent yield reduction, to over $20 billion (0.13 per cent) under a 40 per cent reduction. Under the worst case scenarios welfare decreases by 0.2 per cent and 0.34 per cent of GDP under 10 per cent and 40 per cent reductions of yield. Under scenario A the average global increase in overall production costs could thus be small (perhaps a few per cent of world agricultural GDP). But under the worst case scenario the average costs of world agricultural production could amount to over 18 per cent of world agricultural GDP (see Chapter 7).

It should be emphasized, once again, that these analyses do not take account of changes in technology and management (such as changes in land allocations to different crops) that would almost certainly accompany any alteration of productive potential due to climatic change. Being a static model the SWOPSIM analyses are "switch-on" experiments that, quite unrealistically, assume that the changes in yield occur now, under present-day conditions of technology, land use and trading agreements, rather than over several decades.

However, these preliminary results serve to test the sensitivity of the world food system to changes of climate, indicating what magnitudes and rates of climatic change could be absorbed without severe impact and, alternatively, what magnitudes and rates could seriously perturb the system. The indications are that yield reductions of up to 20 per cent in the major mid-latitude grain-exporting regions could be tolerated without a major interruption of global food supplies. However, the increase in food prices (7 per cent under a 10 per cent yield decrease) could seriously influence the ability of food-deficit countries to pay for food imports, eroding the amount of foreign currency available for promoting development of their non-agricultural sectors.

8. ADAPTING TO CLIMATIC CHANGE

Our assessment of possible effects has, up to this point, assumed that technology and management in agriculture do not alter significantly in response to climatic change, and thus do not alter the magnitude and nature of the impacts that may stem from that change. It is certain, however, that agriculture will adjust and, although these adjustments will be constrained by economic and political factors, it is likely that they will have an important bearing on future impacts. Two broad types of adjustment may be anticipated: changes in land use and changes in management.

CHANGES IN LAND USE

Three types of land-use change will probably have the greatest effect: changes in farmed area, crop type and crop location.

Changes in farmed area

Where warming tends to reduce climatic constraints on agriculture, such as in high-latitude and high-altitude areas, an extension of the farmed area can be expected if other environmental factors and economic incentives permit. Expansion may be most marked in the USSR and northern Europe, where terrain and soils will permit further reclamation.[1, 2] But it may be more limited by inappropriate soils in much of Canada, with the exception of the Peace River region in northern Alberta and parts of Ontario.[3, 4] There may also be potential for high-latitude reclamation in some of the valleys of central Alaska, in northern Japan and in southern Argentina and New Zealand.[5, 6]

Warming may also tend to induce an upward extension of the farmed area in upland regions. For example, in the European Alps a 1°C warming can be expected to raise climatic limits to cultivation by about 150 m. Similar upward shifts are estimated to increase the farmed area significantly in high mid-latitude mountain environments such as northern Japan and South Island New Zealand.[7, 8] These shifts of the limit of the farmed area imply major impacts on the semi-natural environment and on extensive rangeland economies in mountain regions, such as Alpine pastures, which may come under pressure both from the upward advance of more intensive agriculture and from afforestation.

In regions where reduced moisture availability leads to decreased productive potential, particularly where agriculture is at present only marginally productive, there may occur a significant decline in acreage under use. This may occur, for example, in parts of the eastern Mediterranean if projected decreases in rainfall are correct, and also possibly in western Australia.[9, 10] In the south-east USA increased heat stress and evaporation losses may reduce profitability to the point where commercial cropping becomes non-viable. For example, the cropped acreage in the southern Great Plains of the USA is estimated to decline by between 5 per cent and 23 per cent under a warmer and drier $2 \times CO_2$ climate (+3.8 to +6.3°C, soil moisture −10 per cent). This may be partially compensated for by increases in cultivated area in the Great Lakes region.[11]

Changes in crop type

Changes to crops with higher thermal requirements

In regions where there are substantial increases in the warmth of the growing season (and where output is currently limited by temperature rather than by rainfall) it is logical that substitution by crops with higher thermal requirements, that would make fuller use of the extended and more intense growing season, should allow higher yields. Recent impact assessments have considered this as a predictable response in the USA, UK, Japan and New Zealand.[11, 12, 7, 6]

An illustration of the potential for this form of adaptation is given in Figure 8.1 which indicates the different levels of rice yield that could be achieved in Hokkaido (northern Japan) from different rice varieties. Yields of present-day quick-maturing rice varieties would increase by about 4 per cent with a 35 per cent increase in summer warmth (consistent with the GISS $2 \times CO_2$ climate). However, the adoption of a late-maturing rice variety (at present grown in central Japan) would enable greater advantage to be taken of the warmer climate, with yields increased by about 25 per cent.[7]

Figure 8.1 Simulated year-to-year variations in rice yield under observed (present-day) climate and GISS 2 × CO_2 climate (mean annual temperature +3.5°C, precipitation +8%) for the period 1974–83 in Hokkaido (N. Japan). Estimates are for current technology and adjusted technology. (*Source:* Yoshino, *et al.*, 1988).[7]

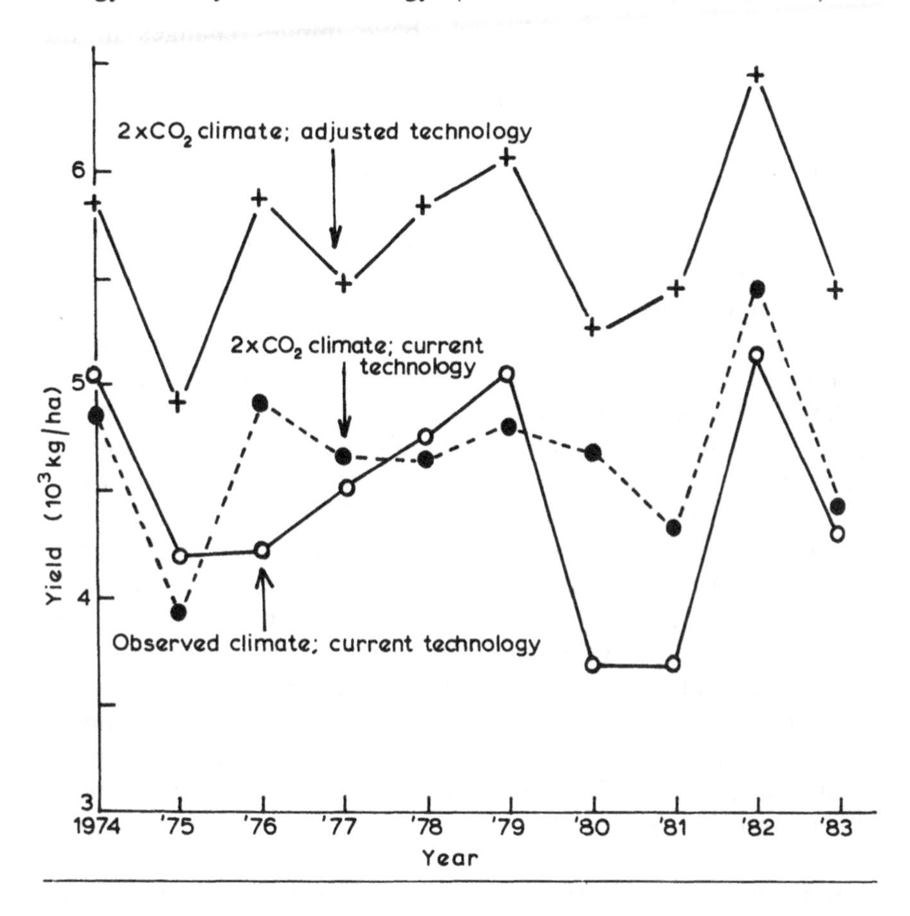

Changes to more drought-tolerant crops

Where moisture rather than temperature is more generally the current climatic constraint on output, or where increases in temperature could well lead to higher rates of evapotranspiration and thus to reduced levels of available moisture, there may occur a switch to crops with lower moisture requirements. Once again the lack of information on likely changes in rainfall makes further speculation on this unprofitable, particularly at lower latitudes. However, there is some evidence that, at high mid-latitudes, a switch from spring to winter varieties of cereals would be one strategy for avoiding losses resulting from more frequent dry spells in the early summer. This might be the case in Scandinavia and on the Canadian prairies.[13, 14]

Changes in crop location

The switch of crops considered above implies changes in the allocation of land to different uses. In general, land uses that show a greater increase in productivity than others are likely to increase their comparative advantage over competing uses; and, given sufficient change in the pattern of comparative advantage, decisions then may follow which involve a change in use. The amount of land-use change is likely to depend on how finely land uses in a given area are currently tuned to economics and climate, and much will depend on the changes in price that are largely determined by changes in potential in other areas. The response is therefore likely to be complex and extremely difficult to predict.

An illustration of potential adjustments of land use in response to possible climatic change is given in Figure 8.2. This indicates the differential yield effects for a range of crops of a 1°C increase in mean annual temperature in the Central Region around Moscow. An economic optimizing model simulated the altered land use that would be likely to maximize output and minimize production costs. Under a 1°C warming, yields of winter wheat and silage maize may increase because they are currently limited by temperature, but yields of temperate-zone crops such as barley and oats are reduced. Experiments suggest that the optimal reallocation of land under these circumstances would be a 30 per cent and 5 per cent increase in land under winter wheat and maize, and a 20–30 per cent decrease in land under barley and oats.[15, 1]

The broadscale changes in crop location imply a general poleward shift of present-day agricultural zones. This is likely to be most pronounced in mid- and high latitudes partly because warming will be most marked here, but largely because it is in these regions that latitudinal zoning is most evident as a result of differences in available warmth for crop maturation.

An illustration of the possible extent of such shifts was given in Figure 5.6 (Chapter 5) relating to the northern limit of maize production in the UK. This is at present located in the extreme south of the country, but would shift about 300 km northwards for each °C rise in mean annual temperature.[16] Broadly similar shifts

Figure 8.2 Effects of adjustments to crop allocation in the Central Region (northern USSR) on agricultural receipts and production costs. (*Source:* Parry and Carter 1988).[15]

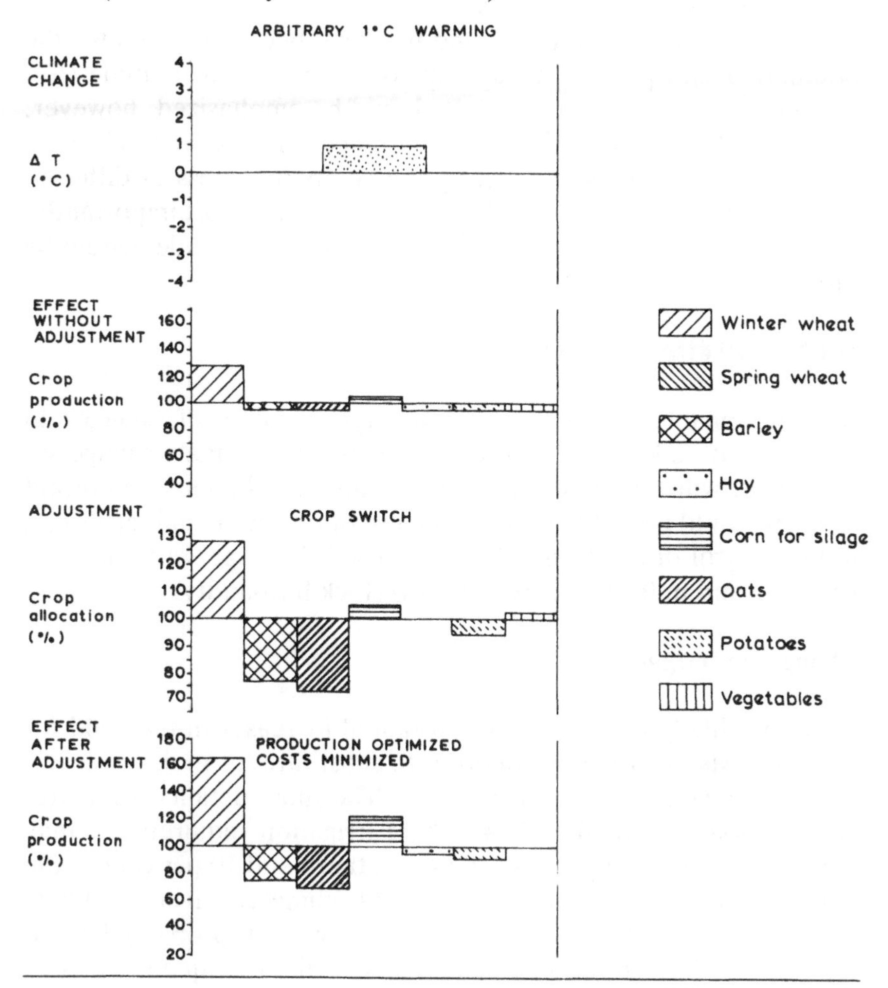

oilseeds and a wider range of fruits and vegetables would be viable.[13] In southern Europe higher temperatures imply a more northerly location of present limits of citrus, olives and vines.[17]

In central North America, zones of farming types are estimated to shift about 175 km northwards for each °C of warming, resulting in reduced intensity of use in the south, and increased intensity in the north where soils and terrain permit.[11] In Canada up to four million hectares of currently unused northern soils may be suitable for cultivation in a warmer climate.[18] A northward movement of production would suggest that a sizeable area of output currently located in the northern Great Plains of the USA would relocate in the southern Canadian prairies.

Similar southward shifts of land use have been suggested for the Southern Hemisphere, perhaps up to six degrees of latitude (670 km) under a 3°–4°C warming.[6] It should be emphasized, however, that these broadscale effects will be much affected at local levels by regional variations in soils, by the competitiveness of different crops and their environmental requirements and, most importantly, by regional patterns of rainfall, none of which can adequately be projected at the present.

CHANGES IN MANAGEMENT

It is reasonable to expect that a large number of changes in management, adopted over time as the effects of climate change are perceived, will modify effects outlined above. The most important changes would probably occur in the use of irrigation and fertilizers, in the control of pests and diseases, in soil drainage, in farm infrastructure and in forms of crop and livestock husbandry.

Changes in irrigation

There are likely to occur very substantial increases in the need for and the costs of irrigation, in order to substitute for moisture losses due to increased evapotranspiration. The most detailed estimates yet available are for the USA, where irrigation requirements may increase by about 25 per cent in the southern and 10 per cent in the northern Great Plains under a 2 × CO_2 climate. Given the likely increased rate of groundwater depletion, this will probably lead to significantly higher costs of production, with consequent shifts to less water-demanding uses in the most affected areas.[11]

Substantially increased irrigation needs are also projected for most of western and southern Europe.[17] Elsewhere, although data are not available, it is probable that effects will be similar where available water is reduced. Where there are substantial rainfall increases, changes in management may be needed to tackle soil erosion, etc. (see below). Irrigation is practised mostly in arid or semi-arid regions where there is already a strain on available resources. Most of the irrigated land of the world is in Asia with a rapidly-increasing population and not much latitude for increases in use of land and water resources. To counter the increased water demand due to climate change, tighter water-management practices should lead to higher irrigation efficiency.

Changes in fertilizer use

More use of fertilizers may be needed to maintain soil fertility where increases in leaching result from increased rainfall. In other regions, warming may increase productive potential to the extent that current levels of output can be achieved with substantially lower amounts of fertilizers. In Iceland, for example, fertilizer use could possibly be halved under a warming of +4°C while maintaining present-day output.[19] This might provide significant savings in costs if fertilizer prices rise as a result of rising energy costs designed to reduce GHG emissions rates.

Much will depend on other factors, for example to what extent higher CO_2 levels will make nutrients more limiting, thus requiring more use of fertilizers, and how future changes in energy prices affect the cost of fertilizers.

Control of pests and diseases

The costs of these are likely to alter substantially, although it is quite impossible to specify them with any degree of detail. Possibly most important for global cereal production may be the costs of controlling the spread of subtropical weed species into current major cereal-producing regions.[11]

Soil drainage and the control of erosion

Adjustments in management are likely to be necessary in tropical regions, particularly those characterized by monsoon rainfall, where there may be an overall increase in rainfall receipt, and possibly, an increase in the intensity of rainfall. Recent assessments (for example, in the USSR) have indicated that, over the longer term, reduced soil fertility, increased salinity and the costs of erosion control may more than offset the beneficial effects of a warmer climate, leading ultimately to reduced yields and higher production costs.[1]

Changes in farm infrastructure

Regional shifts of farming types and altered irrigation requirements imply major changes in capital equipment, in farm layout and in agricultural support services (marketing, credit, etc.). In the USA it has been estimated that these will be substantial.[11] Because of the very large costs involved only small, incremental adjustments may occur without changes in government policies.

Changes in crop and livestock husbandry

The adjustments summarized above imply a plethora of small, but important changes in farm husbandry. In particular there are likely to occur very many alterations to the timing of various farm operations such as tillage (ploughing, sowing, harvesting, etc.), fertilizing and pest and weed control (spraying, etc.), because the timing of these in the present farming calendar, though of course different for various parts of the world, is frequently affected by present climate. Particular aspects of husbandry are also likely to be affected, such as the density of planting, the use of fallowing and mulching and the extent of inter-cropping. These aspects are, today, frequently part of a package of strategies designed to mitigate the adverse effects (and exploit the beneficial effects) of present-day climate. Thus a change of climate implies a re-tuning of these strategies to harmonize with the new set of climatic conditions.

9. CONCLUSIONS

EFFECTS ON FOOD SECURITY

The conclusions of this book are similar to those of the recent IPCC assessment (by the same author) of potential effects of climate change on agriculture.[1] It has been the purpose of the book to consider, in more detail than could be covered by the IPCC report, the reasoning behind these conclusions, their implications for food security and the most appropriate courses of action.

On balance, the evidence is that food production at the global level can, in the face of estimated changes of climate, be maintained at essentially the same level as would have occurred without climate change, but the cost of achieving this is unclear. It could be very large.

Moreover, there may well occur severe negative impacts of climatic change on food supply at the regional level, particularly in regions of high present-day vulnerability least able to adjust technically to such effects. Increases in productive potential at high mid-latitudes and high latitudes, while being of regional importance, are not likely to open up large new areas for production. The gains in productive potential here due to climatic warming would be unlikely to balance possible large-scale reductions in potential in some major grain-exporting regions at mid-latitude.

MAGNITUDES OF POSSIBLE DISLOCATION

From the estimate of changes in productive potential for the changes of climate outlined in this book, the cost of producing some mid-latitude crops such as maize and soybean could increase, reflecting a small net decrease in the global food production-capability of these crops. Rice production could, however, increase if available moisture increased in south-east Asia, although these effects may be limited by increased cloudiness and temperature. The average global increase in overall production costs could thus be small (perhaps a few per cent of world agricultural GDP). Much depends however, on how beneficial are the so-called "direct" effects of increased CO_2 on crop yield. If plant productivity is substantially enhanced and more moisture is available in some major production areas, then world productive potential of staple cereals could increase relative to demand. If, on the contrary, there is little beneficial direct CO_2 effect *and* climate changes are negative for agricultural potential in all or most of the major food-exporting areas, then the average costs of world agricultural production could increase significantly. These increased costs could amount to over 10 per cent of world agricultural GDP.

THE MOST VULNERABLE REGIONS AND SECTORS

On the basis both of limited resource capacity in relation to present-day population and of possible future diminution of the agricultural resource base as a consequence of reduced crop water availability, two broad sets of regions appear most vulnerable to climatic change: a) some semi-arid tropical and subtropical regions (possibly western Arabia, the Maghreb, western West Africa, Horn of Africa and Southern Africa, eastern Brazil) and b) some humid tropical and equatorial regions (possibly south-east Asia, Central America).

In addition, certain regions that are currently net exporters of cereals could also be characterized by reduced crop-water availability and thus by reduced productive potential as a result of climatic changes. Any decrease in production in these regions could markedly affect future global food prices and patterns of trade. These regions include, for example: southern Europe, southern USA, parts of South America, western Australia.

THE EFFECTS OF ALTERED CLIMATIC EXTREMES

Relatively small changes in the mean values of rainfall and temperature can have a marked effect on the frequency of extreme levels of available warmth and moisture. For example, the number of very hot days which can cause damaging heat stress to temperate crops and livestock could increase significantly in some regions as a result of a 1° or 2°C increase in mean annual temperatures. Similarly, reductions in average levels of soil moisture as a result of higher rates of evapotranspiration could increase substantially the number of days below a minimum threshold of water availability for given crops.

Although we know little, at present, about how these frequencies of extreme events may alter as a result of climatic change, the potential impact of concurrent drought or heat stress in the major food-exporting regions of the world could be severe. In addition, relatively small decreases in rainfall, changes in rainfall distribution or increases in evapotranspiration could markedly increase the probability, intensity and duration of drought in currently drought-prone (and often food-deficient) regions. Increase in drought-risk represents potentially the most serious impact of climatic change on agriculture both at the regional and the global level.

EFFECTS ON CROP GROWTH POTENTIAL, LAND DEGRADATION, PESTS AND DISEASES

Higher levels of atmospheric CO_2 are expected to enhance the growth rate of some staple cereal crops such as wheat and rice, but not of others such as millet, sorghum and maize. The use of water by crop plants may also be more efficient under higher CO_2 levels. However, it is not clear how far these potentially beneficial "direct" effects of enhanced atmospheric CO_2 will be manifested in the farmer's field rather than in the experimental glasshouse.

Warming is likely to result in a poleward shift of thermal limits of agriculture, which may increase productive potential in high-latitude regions. But soils and terrain may not enable much of this potential to be realized. Moreover, shifts of moisture limits in some semi-arid and subhumid regions could lead to significant reductions of potential here, with serious implications for regional food supplies in some developing countries.

Temperature increases may extend the geographic range of some insect pests, diseases and weeds, allowing (for example) their expansion to new regions as they warm and become suitable habitats. Changes in temperature and precipitation may also influence soil characteristics.

REGIONAL IMPACTS

Impacts on potential yields are likely to vary greatly according to types of climatic change and types of agriculture. In the northern mid-latitude regions where summer drying may reduce productive potential (e.g. in the south and central USA and in southern Europe) studies have estimated yield potential to fall by 10–30 per cent under an equilibrium $2 \times CO_2$ climate. Towards the northern edge of current core-producing regions, however, warming may enhance productive potential in climatic terms. When combined with direct CO_2 effects, increased climatic potential could be substantial, though in actuality it may be limited by soils, terrain and land use.

There are indications that warming could lead to an overall reduction of cereal production potential in North America, and to reduced potential in southern Europe but increased potential in northern Europe. Warming could allow increased agricultural output in regions near the northern limit of current production in the USSR, but output in the Ukraine and Kazakhstan could only increase if corresponding increases in soil moisture were to occur, and this is at present uncertain.

Little is known about likely impacts in semi-arid and humid tropical regions, because production potential here largely depends on crop-water availability and the regional pattern of possible changes in precipitation is unclear at present. It is prudent, however, to assume that crop-water availability could decrease in some regions. Under these circumstances there could be substantial regional dislocation of access to food supply.

ADAPTATION IN AGRICULTURE

In some parts of the world, climatic limits to agriculture are estimated to shift by 200–300 km per °C of warming (or 100 km per decade under the IPCC Business-As-Usual scenario). The warming-induced upward shift in thermal zones along mountain slopes would be in the order of 150–200 m per °C of warming.

Agriculture has an ability to adjust, within given economic and technological constraints, to a limited rate of climatic change. This capability probably varies greatly between regions and sectors, but no thorough analysis of adaptive capacity has yet been conducted for the agricultural sector.

In some currently highly variable climates farmers may be more adaptable than those in regions of more equable climate. But in less developed economies, and particularly in some marginal types of agriculture, this intrinsic adaptive capability may be much less. It is important to establish in more detail the nature of this adaptability, and thus help determine critical rates of climatic change that would exceed those that could be accommodated by within-system adjustments.

FUTURE TASKS

This book has emphasized the inadequacy of our present knowledge. It is clear that more information on potential impacts would help us identify the full range of potentially useful responses and assist in determining which of these may be most valuable. Some priorities for future research may be summarized as follows.

- Improved knowledge is needed of effects of changes in climate on crop yields and livestock productivity in different regions and under varying types of management.

- Improved understanding is needed of the effects of changes in climate on other physical processes, for example on rates of soil erosion and salinization; on soil nutrient depletion; on pests, diseases and soil microbes, and their vectors; on hydrological conditions as they effect irrigation water availability.

- An improved ability is required to "scale up" our understanding of effects on crops and livestock to effects on farm production, on village production, and on national and global food supply. This is particularly important because policies must be designed to respond to impacts at national and global levels.

- Further information is needed on the effects of changes in climate on social and economic conditions in rural areas (e.g., employment and income, equity considerations, farm infrastructure and support services).

- Further information is needed on the range of potentially effective technical adjustments at the farm and village level (e.g., irrigation, crop selection, fertilizing, etc.) and on the economic, environmental and political constraints on such adjustments. In particular, national and international centres of agricultural research should consider the potential value of new research programmes aimed at identifying or developing cultivars and management practices appropriate for altered climates.

- Finally, information is needed on the range of potentially effective policy responses at regional, national and international levels (e.g., reallocations of land use, plant breeding, improved agricultural extension schemes, large-scale water transfers, etc.).

To date, less than a dozen detailed regional studies have been completed that serve to assess the potential impact of climatic changes on agriculture. It should be a cause for concern that we do not, at present, know whether changes of climate are likely to increase the overall productive potential for global agriculture, or to decrease it. There is therefore currently no adequate basis for *predicting* likely effects on food production at the regional or world scale. All that is possible at present is informed speculation. The risks that stem from such levels of ignorance are great. A comprehensive, international research effort is required, now, to redeem the situation.

Further reading

On the Greenhouse Effect and climatic change: most up-to-date is the IPCC report, *Scientific Assessment of Climate Change* (Geneva and Nairobi: World Meteorological Organisation and United Nations Environment Programme, 1990). An overall review can be found in Bolin, B., Döös, B.R., Jäger, J. and Warrick, R.A., *The Greenhouse Effect, Climatic Change and Ecosystems*, SCOPE 29 (Chichester: John Wiley and Sons, 1986).

On methods of assessing possible impacts: a comprehensive survey is Kates, R.W., Ausubel, J.H. and Berberian, M. (eds) *Climate Impact Assessment*, SCOPE 27 (Chichester: John Wiley and Sons, 1985).

On potential (general) impacts of climate change: the literature is sparse, but see IPCC, *The Potential Impacts of Climate Change* (Geneva and Nairobi: World Meteorological Organisation and United Nations Environment Programme, 1990). See also: Rosenberg, N.J., Easterling, W.E., Crosson, P.R. and Darmstadter, J. (eds) *Greenhouse Warming: Abatement and Adaptation*, RFF Proceedings (Washington, DC: Resources for the Future, 1989).

On potential impacts on agriculture: See Parry, M.L., Carter, T.R. and Konijn, N.T. (eds) *The Impact of Climatic Variations on Agriculture, Volume 1, Assessments in Cool Temperate and Cold Regions. Volume 2, Assessments in Semi-Arid Regions* (Dordrecht, The Netherlands: Kluwer, 1988). There is no comparable study of effects in the humid tropics.

Comprehensive national assessments of effects on agriculture have been conducted for the USA, Canada, Iceland, Finland, the USSR and Japan. Full references to these are given in the introduction to Chapter 6. At the time of writing (1990) four other countries are covered by reviews of existing knowledge rather than by new assessments. These are: Australia, New Zealand, UK and West Germany. Reference to these will also be found in Chapter 6.

References and Notes

Chapter 1

1. Working Group II of the IPCC considered impacts on agriculture and forestry, natural ecosystems, hydrology and water resources, industry, transport and human settlement, oceans and coastal zones, and regions covered by snow and ice (the cryosphere). The report on agriculture drew upon the work of more than 100 scientists in 50 different countries. The lead author was M.L. Parry. Specific sections covered impacts from the "direct" effects of CO_2 (by J.I.L. Morison), effects on pests (J.H. Porter), effects on prices and trade (J. Reilly) and effects due to sea-level rise (L.J. Wright). *See*: Intergovernmental Panel on Climate Change, *The Potential Impacts of Climate Change: Impacts on Agriculture and Forestry* (Geneva and Nairobi: World Meteorological Organisation and United Nations Environment Programme, 1990).
2. Intergovernmental Panel on Climate Change, *Scientific Assessment of Climate Change: Policymakers Summary* (Geneva and Nairobi: World Meteorological Organisation and United Nations Environment Programme, 1990).
3. Bolin, B., Döös, B.R., Jäger, J., and Warrick, R.A., *The Greenhouse Effect, Climatic Change and Ecosystems*, SCOPE 29 (Chichester: John Wiley & Sons, 1986).
4. Parry, M.L., "The potential impact on agriculture of the 'Greenhouse Effect' ", in Bennett, R.M. (ed.), *The 'Greenhouse Effect' and UK Agriculture*, CAS Paper 19 (Reading: Centre for Agricultural Strategy, 1989), pp. 27-46. Parry, M.L. "The potential impact on agriculture of the Greenhouse Effect", *Land Use Policy* vol. 7: 109-123 (1990).
5. Schneider, S.H. and Rosenberg, N.J., "The Greenhouse Effect: its causes, possible impacts, and associated uncertainties", in Rosenberg, N.J., Easterling, W.E. III, Crosson, P.R., and Darmstadter, J. (eds.), *Greenhouse Warming: Abatement and Adaptation*, RFF Proceedings (Washington, DC: Resources for the Future, 1989), pp.7-34.
6. FAO, *Land, Food and People* (Rome: UN Food and Agriculture Organisation, 1984).
7. FAO, *Trade Yearbook*, vol. 42 (Rome: UN Food and Agriculture Organisation, 1988).
8. International Wheat Council, *Market Report* (London: International

Wheat Council, 10 July, 1989).
9. Parry, M.L., *Climatic Change, Agriculture and Settlement* (Folkestone: Wm Dawson & Sons, 1978).
10. Parry, M.L., "The impact of climatic variations on agricultural margins", in Kates, R.W., Ausubel, J.H., and Berberian, H., (eds), *Climate Impact Assessment*, SCOPE 27 (Chichester: John Wiley and Sons, 1985), pp. 351-68.
11. Baird, A., O'Keefe, P., Westgate, K. and Wisner, B., *Towards an Explanation and Reduction of Disaster Proneness*, Occasional Paper 11 (Bradford, UK: University of Bradford, Disaster Research Unit, 1975).

Chapter 2

1. Intergovernmental Panel on Climate Change, *Scientific Assessment of Climate Change: Policymakers Summary* (Geneva and Nairobi: World Meteorological Organisation and United Nations Environment Programme, 1990).
2. Watson, R.T., Rodhe, H., Oeschger, H. and Sieganthaler, U. "Greenhouse gases and aerosols", in *IPCC Scientific Assessment of Climate Change* (Geneva and Nairobi: World Meteorological Organisation and United Nations Environment Programme, 1990).
3. Santer, B.D., Wigley, T.M.L., Schlesinger, M.E. and Mitchell, J.F.B. *Developing climate scenarios from equilibrium GCM results*, Max-Planck-Institute für Meteorologie Report No. 47 (Hamburg: Max-Planck-Institute, 1990).
4. Mitchell, J.F.B., Manabe, S., Meleshko, V., Tokioka, T. "Equilibrium climate change and its implications for the future," in *IPCC Scientific Assessment of Climate Change* (Geneva and Nairobi: World Meteorological Organisation and United Nations Environment Programme 1990).
5. FAO, *Trade Yearbook*, vol. 42 (Rome: UN Food and Agriculture Organisation, 1988).
6. In addition to the three high resolution models shown in Figure 2.4, the models are Geophysical Fluid Dynamics Laboratory (GFDL), Goddard Institute for Space Studies (GISS) and National Center for Atmospheric Research (NCAR). Kellogg, W.W. and Zhao, Z., "Sensitivity of soil moisture to doubling of carbon dioxide in climate model experiments. Part I: North America", *Journal of Climate*, vol. 1: pp. 348-66 (1988).
7. Zhao, Z., and Kellogg, W.W., "Sensitivity of soil moisture to doubling carbon dioxide in climate model experiments. Part II: The Asian monsoon regions", *Journal of Climate*, vol. 1: pp. 367-78 (1988).
8. Parry, M.L. *Climatic Change, Agriculture and Settlement* (Folkestone: Wm Dawson & Sons, 1978).
9. Parry, M.L., "The impact of climatic variations on agricultural margins", in Kates, R.W., Ausubel, J.H., and Berberian, H., (eds), *Climate Impact Assessment*, SCOPE 27 (Chichester: John Wiley and

Sons, 1985), pp. 351-68. Also see: Parry, M.L. and Carter, T.R., "The effect of climatic variations on agricultural risk", *Climatic Change*, vol. 7, pp. 95-110 (1985).

10. Wigley, T.M.L., "Impact of extreme events", *Nature* Vol. 316: pp. 106-7 (1985).
11. Mearns, L.O., Katz, R.W., and Schneider, S.H., "Extreme high-temperature events: changes in their probabilities with changes in mean temperature", *Journal of Climate and Applied Meteorology*, vol. 23: pp. 1601-13 (1984).

Chapter 3

1. Kates, R.W., Ausubel, J.H. and Berberian, M. (eds), *Climate Impact Assessment*, SCOPE 27 (Chichester: John Wiley and Sons, 1985).
2. Climate Impact Assessment Program (CIAP), *Impacts of Climatic Change on the Biosphere*, Monograph 5 (Washington, DC: US Department of Transportation, 1975).
3. National Defense University (NDU), *Crop Yields and Climatic Change to the Year 2000*, vol.1 (Washington, DC: Fort Lesley J. McNair, 1980).
4. Parry, M.L. and Carter, T.R., "The assessment of the effects of climatic variations on agriculture: aims, methods and summary of results", in Parry, M.L., Carter, T.R., and Konijn, N.T. (eds), *The Impact of Climatic Variations on Agriculture, Volume 1, Assessments in Cool Temperate and Cold Regions* (Dordrecht, The Netherlands: Kluwer, 1988), pp. 11-96.
5. Garcia, R., *Drought and Man: The 1972 Case History. Vol. 1: Nature Pleads Not Guilty* (New York: Pergamon Press, 1981).
6. Meinl, H,. Bach, W., Jäger, J., Jung, H.-J., Knottenberg, H., Marr, G., Santer, B., and Schwieren, G., *Socio-economic Impacts of Climatic Changes Due to a Doubling of Atmospheric CO_2 Content* (Brussels: Commission of the European Communities, Contract No. CL1-063-D, 1984).
7. Callaway, J.M., Cronin, F.J., Currie, J.W. and Tawil, J., *An Analysis of Methods and Models for Assessing the Direct and Indirect Impacts of CO_2-induced Environmental Changes in the Agricultural Sector of the US Economy*, PNL-4384 (Richland, Washington: Pacific Northwest Laboratory, Battelle Memorial Institute, 1982).
8. Parry, M.L., Carter, T.R. and Konijn, N.T. (eds), *The Impact of Climatic Variations on Agriculture, Volume 1, Assessments in Cool Temperate and Cold Regions* (Dordrecht, The Netherlands: Kluwer, 1988).
9. Parry, M.L., Carter, T.R. and Konijn, N.T. (eds), *The Impact of Climatic Variations on Agriculture, Volume 2, Assessments in Semi-Arid Regions* (Dordrecht, The Netherlands: Kluwer, 1988).
10. Thornthwaite, C.W., "The climates of North America according to a new classification", *Geographical Review*, vol. 21: 633-55 (1931).

11. Palmer, W.C., *Meteorological Drought* (Washington, DC: US Department of Commerce, Research Paper No. 45, 1965).
12. Carter, T.R., Konijn, N.T. and Watts, R.G., "The choice of first-order impact models", in Parry, M.L., Carter, T.R. and Konijn, N.T. (eds), *The Impact of Climatic Variations on Agriculture, Volume 1, Assessments in Cool Temperate and Cold Regions* (Dordrecht, The Netherlands: Kluwer, 1988), pp.97-124.
13. For reviews of these see: Baier, W., "Crop-weather models and their use in yield assessments", *WMO Technical Note No. 151*, (Geneva: World Meteorological Organisation, 1977); Baier, W., "Agroclimatic modeling: an overview", in Cusack, D.F. (ed.), *Agroclimatic Information for Development: Reviving the Green Revolution* (Boulder, Colorado: Westview, 1982); Nix, H.A., "Agriculture", in Kates, R.W., Ausubel, J.H., and Berberian, M. (eds), *Climate Impact Assessment*, SCOPE 27 (Chichester: John Wiley and Sons, 1985) pp. 105-30; and Carter, *et al.*, note 12.
14. Carter, T.R., Parry, M.L., and Porter, J.H.,"Climatic change and future agroclimatic potential in Europe" (forthcoming in *International Journal of Climatology*, 1990).
15. See, for example, Parry, *et al.* 1988, note 8.

Chapter 4

1. Pearch, R.W. and Bjorkman, O., "Physiological effects", in Lemon, E.R. (ed.), *CO_2 and Plants: The Response of Plants to Rising Levels of Atmospheric CO_2* (Boulder, Colorado: Westview Press, 1983), pp. 65-105.
2. Acock, B., and Allen, L.H. Jr, "Crop responses to elevated carbon dioxide concentrations", in Strain, B.R., and Cure, J.D. (eds), *Direct Effects of Increasing Carbon Dioxide on Vegetation*, DOE/ER-0238 (Washington, DC: US Dept. of Energy, 1985).
3. Hillel, D., and Rosenzweig, C., *The Greenhouse Effect and Its Implications Regarding Global Agriculture*, Research Bulletin No. 724 (Amherst, Massachusetts: Massachusetts Agricultural Experiment Station, April, 1989).
4. Akita, S., and Moss, D.N., "Photosynthetic responses to CO_2 and light by maize and wheat leaves adjusted for constant stomatal apertures", *Crop Science*, vol. 13: pp. 234-237 (1973).
5. Morison, J.I.L., "Plant growth in increased atmospheric CO_2", in Fantechi, R. and Ghazi, A. (eds), *Carbon Dioxide and Other Greenhouse Gases: Climatic and Associated Impacts* (Dordrecht, The Netherlands: CEC, Reidel, 1989), pp. 228-244.
6. Edwards, G.E., and Walker, D.A., *C3,C4: Mechanisms and Cellular and Environmental Regulation of Photosynthesis* (Oxford: Blackwell, 1983).
7. Warrick, R.A. and Gifford, R. with Parry, M.L., "CO_2, climatic change and agriculture", in Bolin, B., Döös, B.R., Jäger, J., and Warrick, R.A. (eds), *The Greenhouse Effect, Climatic Change and Ecosystems*, SCOPE 29 (Chichester: John Wiley and Sons, 1986),

pp. 393-473.
8. Morison, J.I.L., "Intercellular CO_2 concentration and stomatal response to CO_2", in Zeiger, E., Cowan, I.R. and Farquhar, G.D. (eds), *Stomatal Function* (Stanford: Stanford University Press, 1987), pp. 229-251.
9. Cure, J.D. and Acock, B., "Crop responses to carbon dioxide doubling: a literature survey", *Agricultural and Forest Meteorology*, Vol. 38: pp. 127-145 (1986).
10. Gifford, R.M., "Direct effect of higher carbon dioxide levels concentrations on vegetation", in Pearman, G.I. (ed.), *Greenhouse: Planning for Climate Change* (Australia: CSIRO 1988), pp. 506-519.
11. Monteith, J.L., "Climatic variation and the growth of crops", *Quarterly Journal of the Royal Meteorological Society,* vol. 107: pp. 749-774 (1981).
12. Squire, G.R. and Unsworth, M.H. "Effects of CO_2 and climatic change on agriculture", *Contract Report to the Department of the Environment* (Sutton Bonnington, UK: Department of Physiology and Environmental Science, University of Nottingham, 1988).
13. Williams, G.D.V., Fautley, R.A., Jones, K.H., Stewart, R.B. and Wheaton, E.E., "Estimating effects of climatic change on agriculture in Saskatchewan, Canada", in Parry, M.L., Carter, T.R. and Konijn, N.T. (eds.) *The Impact of Climatic Variations on Agriculture, Volume 1, Assessments in Cool Temperate and Cold Regions* (Dordrecht, The Netherlands: Kluwer, 1988), pp. 221-379.
14. Brouwer, F.M., "Determination of broad-scale land use changes by climate and soils", *Working Paper WP-88-007* (Laxenburg, Austria: International Institute for Applied Systems Analysis, 1988).
15. Yoshino, M., Horie, T., Seino, H., Tsujii, H., Uchijima, T. and Uchijima, Z., "The effects of climatic variations on agriculture in Japan", in Parry, M.L., Carter, T.R., and Konijn, N.T. (eds), *The Impact of Climatic Variations on Agriculture, Volume 1, Assessments in Cool Temperate and Cold Regions* (Dordrecht, The Netherlands: Kluwer, 1988), pp. 725-868.
16. Pitovranov, S.E., Iakimets, V., Kiselev, V. I. and Sirotenko, O.D., "The effects of climatic variations on agriculture in the subarctic zone of the USSR", in Parry, M.L., Carter, T.R., and Konijn, N.T. (eds), *The Impact of Climatic Variations on Agriculture, Volume 1, Assessments in Cool Temperate and Cold Regions* (Dordrecht, The Netherlands: Kluwer, 1988), pp.617-722.
17. Kettunen, L., Mukula, J., Pohjonen, V., Rantanen, O., and Varjo, U., "The effects of climatic variations on agriculture in Finland", in Parry, M.L., Carter, T.R., and Konijn, N.T. (eds), *The Impact of Climatic Variations on Agriculture, Volume 1, Assessments in Cool Temperate and Cold Regions* (Dordrecht, The Netherlands: Kluwer, 1988), pp.511-614.
18. Bergthorsson, P., Björnsson, H., Dyrmundsson, O., Gudmundsson, B., Helgadottir, A., and Jonmundsson, J.V., "The effects of climatic variations on agriculture in Iceland", in Parry, M.L., Carter, T.R., and Konijn, N.T., (eds), *The Impact of Climatic Variations on*

Agriculture, Volume 1, Assessments in Cool Temperate and Cold Regions (Dordrecht, The Netherlands: Kluwer, 1988), pp.383-509.

19. See, for example, Smith, J.B., and Tirpak, D., *The Potential Effects of Global Climate Change on the United States*, Report to Congress (Washington, DC: US Environmental Protection Agency, 1989). A summary is given in Adams, R.M. (and 9 others), "Global climate change and US Agriculture" *Nature*, vol. 345: 219-224 (1990).
20. Nikonov, A.A., Petrova, L.N., Stolyarova, H.M., Lebedev, V. Yu, Siptits, S.O., Milyutin, N.N., and Konijn, N.T., "The effects of climatic variations on agriculture in the semi-arid zone of European USSR: A. The Stavropol Territory", in Parry, M.L., Carter, T.R., and Konijn., N.T. (eds), *The Impact of Climatic Variations on Agriculture, Volume 2, Assessments in Semi-Arid Regions* (Dordrecht, The Netherlands: Kluwer, 1988), pp. 587-664.
21. Bianca, W. "The significance of meteorology in animal production", *International Journal of Biometeorology*, Vol. 20: 139-156 (1976).
22. Rowntree, P.R., Callander, B.A., and Cochrane, J., "Modelling climate change and some potential effects on agriculture in the UK", *Journal of the Royal Agricultural Society of England* (November 1989).
23. See, for example, FAO, *Report of the Agro-Ecological Zones Project*, World Resources Report, 48 (Rome, Italy: FAO, 1978). Also, Sombroeck, W.G., Braun, H.M.H., and van der Pauw, B.J.A., *Exploratory Soil Map and Agroclimatic Zone Map of Kenya* (Nairobi: Kenya Soil Survey, 1982).
24. Akong'a, J., Downing, T.E., Konijn, N.T., Mungai, D.N., Muturi, H.R., and Potter, H.L., "The effects of climatic variations on agriculture in central and eastern Kenya", in Parry, M.L.,Carter, T. R., and Konijn, N.T. (eds), *The Impact of Climatic Variations on Agriculture, Volume 2, Assessments in Semi-Arid Regions* (Dordrecht, The Netherlands: Kluwer, 1988), pp.123-270.
25. Robertson, G.W., "Development of simplified agroclimate procedures for assessing temperature effects on crop development", in Slatyer, R.O. (ed.), *Plant Response to Climatic Factors. Proceedings of the Uppsala Symposium, 1970* (Paris: UNESCO, 1973), pp. 327-341.
26. Parry, M.L., "The impact of climatic variations on agricultural margins", in Kates. R.W., Ausubel, J.H., and Berberian, M., (eds), *Climate Impact Assessment*, SCOPE 27 (Chichester: John Wiley and Sons, 1985), pp. 351-368.
27. Wigley, T.M.L., "Impact of extreme events", *Nature*, Vol. 316: 106-107 (1985).
28. Parry, M.L., "The significance of the variability of summer warmth in upland Britain", *Weather*, vol. 31: pp. 212-217 (1976).
29. Thompson, L.M., "Weather variability, climatic change, and grain production", *Science*, vol. 188: 535-541 (1975).
30. McQuigg, J.D., "Climate variability and crop yield in high and low temperature regions", in Bach, J., Pankrath, J., and Schneider, S.H. (eds), *Food-Climate Interactions* (Dordrecht, The Netherlands: D. Reidel, 1981).

31. Ramirez, J.M., and Bauer, A., "Small grains response to growing degree units", *Agronomy Abstracts*, p. 163 (1973).
32. Mearns, L.O., Katz, R.W. and Schneider, S.H., "Changes in the probabilities of extreme high temperature events with changes in global mean temperature", *Journal of Climate and Applied Meterology*, vol. 23: 1601-1613 (1984).
33. Salinger, M.J., "The effects of greenhouse gas warming on forestry and agriculture", Draft report for *WMO Commission of Agrometeorology* (1989).
34. Waggoner, P.E., "Agriculture and a climate changed by more carbon dioxide", in US. National Research Council, *Changing Climate* (Washington, DC: National Academy Press, 1983).
35. I am grateful to Julia Porter for permission to reproduce this section, originally drafted by her for the IPCC review of impacts on agriculture. See, IPCC, *The potential impacts of climate on agriculture* (Geneva and Nairobi: WMO and UNEP, 1990).
36. Porter, J.H., personal communication, 1990.
37. Drummond, R.O., "Economic aspects of ectoparasites of cattle in North America", in *Symposium, The Economic Impact of Parasitism in Cattle*, XXIII World Veterinary Congress, Montreal, 1987.
38. Sutherest, R.W., "The role of models in tick control", in Hughes, K.L. (ed.), *Proceedings of the International Conference on Veterinary Preventive Medicine and Animal Production* (Melbourne, Australia: Australian Veterinary Association, 1987), pp. 32-37.
39. Messenger, G., "Migratory Locusts in New Zealand?", *The Weta*, vol. 11-2: 29 (1988). Hill, M.G., and Dymock, J.J., *Impact of Climate Change: Agricultural/Horticultural Systems* (DSIR Entomology Division, submission to New Zealand Climate Programme, Department of Scientific and Industrial Research, Wellington, New Zealand, 1989).
40. Pedgley, D.E. "Weather and the current desert locust plague", *Weather*, vol. 44-4: 168-171 (1989).
41. Beresford, R.M., and Fullerton, R.A., *Effects of climate change on plant diseases*, DSIR Plant Division Submission to Climate Change Impacts Working Group, May, 1989 (Wellington, New Zealand: Department of Industrial and Scientific Research, 1989).
42. Kauppi, P., and Posch, M., "A case study of the effects of CO_2-induced climatic warming on forest growth and the forest sector: A. Productivity reactions of northern boreal forests", in Parry, M.L., Carter, T.R. and Konijn, N.T. (eds), *The Impact of Climatic Variations on Agriculture, Volume 1, Assessments in Cool Temperate and Cold Regions* (Dordrecht, The Netherlands: Kluwer, 1988), pp.183-195.
43. De Groot, R.S., "Assessment of potential shifts in Europe's natural vegetation due to climatic change and implications for conservation", *Young Scientists' Summer Program 1987: Final Report* (Laxenburg, Austria: International Institute for Applied Systems Analysis, 1987).
44. Warrick, R.A., and Oerlemans, J., "Sea level rise", in: *IPPC Scientific Assessment of Climate Change* (Geneva and Nairobi: WMO and UNEP, 1990).

45. UNEP, "Criteria for assessing vulnerability to sea level rise: a global inventory to high risk areas", United Nations Environment Programme and the Government of the Netherlands, Draft Report, (1989).

Chapter 5

1. Parry, M.L., and Carter, T.R., "An assessment of the effects of climatic change on agriculture", *Climatic Change*, Vol. 15: 95-116 (1989).
2. Parry, M.L., "The impact of climatic variations on agricultural margins", in Kates, R.W., Ausubel, J.H., and Berberian, H. (eds), *Climatic Impact Assessment*, SCOPE 27 (Chichester: John Wiley and Sons, 1985), pp. 351-68.
3. Blasing, T.J., and Solomon, A.M., *Response of North American Corn Belt to Climatic Warming*, DOE/N88-004, (Washington, DC: US Department of Energy, Carbon Dioxide Research Division, 1983).
4. Newman, J.E., "Climate change impacts on the growing season of the North American Corn Belt", *Biometeorology*, vol. 7-2: pp. 128-42 (1980).
5. Rosenzweig, C., "Potential $C0_2$-induced climate effects on North American wheat-producing regions", *Climatic Change*, vol. 7: pp. 367-89 (1985).
6. Carter, T.R., Parry, M.L. and Porter, J.H., "Climatic change and future agroclimatic potential in Europe" (Forthcoming in *International Journal of Climatology*, 1990).
7. Yoshino, M., Horie, T., Seino, H., Tsujii, H., Uchijima, T. and Uchijima, Z., "The effects of climatic variations on agriculture in Japan", in Parry, M.L., Carter, T.R., and Konijn, N.T. (eds), *The Impact of Climatic Variations on Agriculture, Volume 1, Assessments in Cool Temperate and Cold Regions* (Dordrecht, The Netherlands: Kluwer, 1988), pp. 725-878.
8. Salinger, M.J., Williams, W.M., Williams, J.M., and Martin, R.J., (eds), *Carbon Dioxide and Climate Change: Impacts on Agriculture* (New Zealand: New Zealand Meteorological Service; DSIR Grasslands Division; MAFTech, 1990).
9. Hansen, J., Fung, I., Lacis, A., Rind, D., Lebedeff, S., Ruedy, R. and Russell, G., "Global climate changes as forecast by Goddard Institute for Space Studies three-dimensional model", *Journal of Geophysical Research*, Vol. 93-D8: pp. 9341-64 (1988).
10. Parry, M.L., Carter, T.R., and Porter, J.H., "The Greenhouse Effect and the future of UK agriculture", *Journal of the Royal Agricultural Society of England*, November: pp. 120-131 (1989).
11. IPCC. *Scientific Assessment of Climate Change: Policy makers Summary* (Geneva and Nairobi: WMO and UNEP, 1990).
12. Parry, M.L., and Carter, T.R., "The assessments of the effects of climatic variations on agriculture: aims, methods and summary of results", in Parry, M.L., Carter, T.R., and Konijn, N.T. (eds), *The*

Impact of Climatic Variations on Agriculture, Volume 1, Assessments in Cool Temperate and Cold Regions (Dordrecht, The Netherlands: Kluwer, 1988), pp. 11-96.

13. Parry, M. L. "Secular climatic change and marginal land", *Transactions of the Institute of British Geographers*, vol. 64; pp. 1-13 (1975).
14. Parry, M.L., *Climatic Change, Agriculture and Settlement* (Folkestone: Wm Dawson & Sons, 1978).
15. Balteanu, D., Ozenda, P., and Kuhn, M., "Impact analysis of climatic change in the central European mountain ranges", Vol. G, *European Workshop on Interrelated Bioclimatic and Land Use Changes* (Noordwijkerhout, The Netherlands: October, 1987).
16. Bravo, R.E., Canadas Cruz, L., Estrada, W., Hodges, T., Knapp, G., Ravelo, A.C., Planchuelo-Ravelo, A.M., Rovere, O., Salcedo Solis, T. and Yugcha, P.T., "The effects of climatic variations on agriculture in the Central Sierra of Ecuador", in Parry, M.L., Carter, T.R., and Konijn, N.T. (eds), *The Impact of Climatic Variations on Agriculture, Volume 2, Assessments in Semi-Arid Regions* (Dordrecht, The Netherlands; Kluwer, 1988), pp. 383-493.
17. Todorov, A.V., "Sahel: The changing rainfall regime and the 'normals' used for its assessment", *Journal of Climate and Applied Meteorology*, vol. 24-2: pp. 97-110.
18. Pittock, A. B. and Nix, H.A., "The effects of changing climate on Australian biomass production – A preliminary study", *Climatic Change*, vol. 8: pp. 243-55 (1986).
19. Kettunen, L., Mukula, J., Pohjonen, V., Rantanen, O., and Varjo, U., "The effects of climatic variations on agriculture in Finland", in Parry, M.L., Carter, T.R. and Konijn, N.T. (eds.), *The Impact of Climatic Variations on Agriculture, Volume 1, Assessments in Cool Temperate and Cold Regions* (Dordrecht, The Netherlands: Kluwer, 1988), pp. 511-614.
20. For 2 × CO_2 studies, see: Pitovranov, S.E., Iakimets, V., Kiselev, V.I. and Sirotenko, O.D, "The effects of climatic variations on agriculture in the subarctic zone of the USSR", in Parry, M.L., Carter, T.R., and Konijn, N.T. (eds), *The Impact of Climatic Variations on Agriculture, Volume 1, Assessments in Cool Temperate and Cold Regions* (Dordrecht, The Netherlands: Kluwer, 1988), pp. 617-722. For palaeo-analogue studies, see: Intergovernmental Panel on Climate Change, *The Potential Impacts of Climate Change: Impacts on Agriculture and Forestry* (Geneva and Nairobi, WMO and UNEP, 1990).
21. Williams, G.D.V., Fautley, R.A., Jones, K. H.,Stewart, R.B. and Wheaton, E.E., "Estimating effects of climatic change on agriculture in Saskatchewan, Canada", in Parry, M.L., Carter, T.R., and Konijn, N.T. (eds) *The Impact of Climatic Variations on Agriculture, Volume 1, Assessments in Cool Temperate and Cold Regions* (Dordrecht, The Netherlands: Kluwer, 1988), pp. 221-379.
22. Smit, B., Brklacich, M., Stewart, R.B., McBride, R., Brown, M., and Bond, D., "Sensitivity of crop yields and land resource potential to climate change in Ontario", *Climatic Change*, vol. 14: pp. 153-74 (1989).

23. Rosenzweig, C., "Potential effects of climate change on agricultural production in the Great Plains: A simulation study", in Smith, J.B. and Tirpak, D.A., *The potential effects of global climate change on the United States, Appendix C – Agriculture* (Washington DC: US Environmental Protection Agency), pp. 3/1–3/43.
24. Squire, G.R., and Unsworth, M.H., "Effects of CO_2 and climatic change on agriculture", *Contract Report to the Department of the Environment* (Sutton Bonnington, UK: Department of Physiology and Environmental Science, University of Nottingham, 1988). Further information for the UK is given in Unsworth's contribution to the First Report of the UK Climate Change Impacts Review Group which is due to be published in early 1991.
25. Santer, B., "The use of general circulation models in climate impact analysis – a preliminary study of the impacts of a CO_2-induced climatic change on Western European agriculture", *Climatic Change*, Vol. 7: pp. 71-93 (1985).
26. Pittoch, A.B., "Potential impacts of climatic change on agriculture, forestry and land use" (personal communication, 1989), 5 pp.
27. Schulze, R.E., "Hydrological responses to long-term climatic change" (personal communication,1989), 9 pp.
28. Van Diepen, C. A., Van Keulen, H., Penning de Vries, F.W.T., Noy, I.G.A.M., and Goudriaan, J., "Simulated variability of wheat and rice yields in current weather conditions and in future weather when ambient CO_2 has doubled", *Simulation Reports CABO-TT*, 14 (Wageningen, The Netherlands: University of Wageningen, 1987).
29. Zhang, Jia-cheng, "The CO_2 problem in climate and dryness in North China", *Meteorological Monthly*, vol. 15-3: pp.3-8 (1989).
30. Terjung, W.H., Ji, H-Y., Hayes, J.T., O'Rourke, P.A., and Todhunter, P.E., "Actual and potential yield for rainfed and irrigated maize in China", *International Journal of Biometeorology*, vol. 28: pp. 115-35.
31. Akong'a, J., Downing, T.E., Konijn, N.T., Mungai, D.N., Muturi, H.R., and Potter, H.L., "The effects of climatic variations on agriculture in Central and Eastern Kenya", in Parry, M.L., Carter, T.R., and Konijn, N.T., (eds), *The Impact of Climatic Variations on Agriculture, Volume 2, Assessments in Semi-Arid Regions* (Dordrecht, The Netherlands: Kluwer, 1988), pp.123-270.
32. Panturat, S., and Eddy, A., "Some impacts on rice yield from changes in the variance of regional precipitation", *Air Group Interim Report to UNEP* (Birmingham, UK: AIR Group, School of Geography, University of Birmingham, 1989).
33. Berthorsson, P., Björnsson, H., Dyrmundsson, O., Gudmundsson, B., Helgadottir, A., and Jonmundsson, J.V., "The effects of climatic variations on agriculture in Iceland", in Parry, M.L., Carter, T.R., and Konijn, N.T. (eds), *The Impact of Climatic Variations on Agriculture, Volume 1, Assessments in Cool Temperate and Cold Regions* (Dordrecht, The Netherlands: Kluwer, 1988).
34. Hales, J., personal communication (1989).
35. Pearman, G.I. (ed), *Greenhouse: Planning for Climate Change* (Melbourne, Australia: CSIRO, 1988).

Chapter 6

1. Parry, M.L., Carter, T.R. and Konijn, N.T. (eds), *The Impact of Climatic Variations on Agriculture, Volume 1, Assessments in Cool Temperate and Cold Regions* (Dordrecht, The Netherlands: Kluwer, 1988).
2. Smit, B., "Climate warming and Canada's comparative position in agriculture", *Climate Change Digest*, CCD89-01 (Environment Canada, 1989).
3. EPA, *The Potential Effect of Global Climate Change on the United States*. Draft Report to Congress (Washington, DC: US Environmental Protection Agency, 1989). Adams, R.M., and 9 others, "Global climate change and US agriculture", *Nature*, Vol. 345: pp. 219-224 (1990).
4. Pearman, G.I., (ed.), *Greenhouse: Planning for Climate Change* (Melbourne, Australia: CSIRO, 1988).
5. Salinger, M.J., Williams, W.M., Williams, J.M., and Martin, R.J. (eds), *Carbon Dioxide and Climate Change: Impacts on Agriculture* (New Zealand: New Zealand Meteorological Service; DSIR Grasslands Division; MAFTech, 1990).
6. DOE, *Possible Impacts of Climate Change on the Natural Environment in the United Kingdom* (London: UK Department of the Environment, 1988). Further information is given in the First Report of the U.K. Climate Change Impacts Review Group which is due to be published in 1990.
7. Study Commission of Eleventh German Bundestag, *Protecting the Earth's Atmosphere: An International Challenge* (Bonn: Bonn University, 1989).
8. Williams, G.D.V., Fautley, R.A., Jones, K.H., Stewart, R.B., and Wheaton, E.E., "Estimating effects of climatic change on agriculture in Saskatchewan, Canada", in Parry, M.L., Carter, T.R., and Konijn, N.T. (eds), *The Impact of Climatic Variations on Agriculture, Volume 1: Assessments in Cool Temperate and Cold Regions* (Dordrecht, The Netherlands: Kluwer, 1988), pp. 221-379.
9. Parry, M.L., and Carter, T.R., "The assessments of the effects of climatic variations on agriculture: aims, methods and summary of results", in Parry, M.L., Carter, T.R., and Konijn, N.T. (eds), *The Impact of Climatic Variations on Agriculture, Volume 1, Assessments in Cool Temperate and Cold Regions* (Dordrecht, The Netherlands: Kluwer, 1988), pp. 11-96.
10. Smit, B,. "Implications of climatic change for agriculture in Ontario", *Climate Change Digest*, CCD87-02 (Environment Canada, 1987).
11. Yoshino, M., Horie, T., Seino, H., Tsujii, H., Uchijima, T. and Uchijima, Z., "The effects of climatic variations on agriculture in Japan", in Parry, M.L., Carter, T.R. and Konijn, N.T. (eds), *The Impact of Climatic Variations on Agriculture, Volume 1, Assessments*

in Cool Temperate and Cold Regions (Dordrecht, The Netherlands: Kluwer, 1988), pp. 725-868.
12. Pittock, A.B., "The Greenhouse Effect, regional climate change and Australian agriculture", Paper presented to *Australian Society of Agronomy, 5th Agronomy Conference*, Perth, September, 1989.
13. Bergthorsson, P., Björnsson, H., Dyrmundsson, O., Gudmundsson, B., Helgadottir, A. and Jonmundsson, J.V., "The effects of climatic variations on agriculture in Iceland", in Parry, M.L., Carter, T.R., and Konijn, N.T. (eds.), *The Impact of Climatic Variations on Agriculture, Volume 1, Assessments in Cool Temperate and Cold Regions* (Dordrecht, The Netherlands: Kluwer, 1988), pp.383-509.
14. Kettunen, L., Mukula, J., Pohjonen, V., Rantanen, O. and Varjo, U., "The effects of climatic variations on agriculture in Finland", in Parry, M.L., Carter, T.R., and Konijn, N.T. (eds), *The Impact of Climatic Variations on Agriculture, Volume 1, Assessments in Cool Temperate and Cold Regions* (Dordrecht, The Netherlands: Kluwer, 1988) pp. 511-614.
15. Pitovranov, S.E., Iakimets, V., Kiselev, V.I. and Sirotenko, O.D., "The effects of climatic variations on agriculture in the subarctic zone of the USSR", in Parry, M.L., Carter, T.R., and Konijn, N.T. (eds), *The Impact of Climatic Variations on Agriculture, Volume 1, Assessments in Cool Temperate and Cold Regions* (Dordrecht, The Netherlands: Kluwer, 1988), pp. 617-722.
16. Yinnikov, K. Ya. and Groismann, P. Ya, "An empirical model of present-day climatic changes", *Meteorologia i Gidrologia*, Vol. 3, pp. 25-28 (1979) (in Russian). English translation available in: *Soviet Meteorology and Hydrology*, Vol. 3 (1979).

Chapter 7

1. Parry, M.L. and Sinha, S.K., "Food Security", Working Group Report in *Conference Proceedings, The Changing Atmosphere: Implications for Global Security* (Toronto, Canada, 27-30 June, 1988, WMO-No, 170, World Meteorological Organisation, 1988), pp. 321-3.
2. This project ("Climate change and International Agriculture") is being jointly implemented by the Goddard Institute for Space Studies (New York, N.Y.) and the Atmospheric Impacts Research (AIR) Group (University of Birmingham, UK). The project leaders are Cynthia Rosenzweig and Martin Parry. Funding is from the US Environmental Protection Agency.
3. Schelling, T., "Climate change: implications for welfare and policy", in *Changing Climate: Report of the Carbon Dioxide Assessment Committee* (Washington, DC: National Academy of Sciences, 1983).
4. Crosson, P., "Greenhouse warming and climate change: why should we care?", *Food Policy*, Vol. 14-2: pp. 107-18 (1989).
5. Sinha, S.K., Rao, N.H., and Swaminathan, M.S., "Food security in the changing global climate", in *Conference Proceedings, The Changing Atmosphere: Implications for Global Security* (Toronto,

Canada, 27-30 June, 1988, WMO-No. 170, World Meteorological Organisation, 1988), pp. 167-92.

6. Parry, M.L. and Duinker, P.N., "The potential impacts of climatic change on agriculture and forestry", in IPCC *Assessment of the Potential Impacts of Climate Change* (Geneva and Nairobi: WMO and UNEP, 1990).
7. Warrick, R.A., and Gifford, R. with Parry, M.L., "CO_2, climatic change and agriculture", in Bolin, B., Döös, B.R., Jäger, J., and Warrick, R.A.(eds), *The Greenhouse Effect, Climatic Change and Ecosystems*, SCOPE 29 (Chichester: John Wiley and Sons, 1986), pp. 393-473.
8. Liverman, D.M., *The Use of a Simulation Model in Assessing the Impacts of Climate on the World Food System*, NCAR Cooperative Thesis No. 77 (Boulder, Colorado: National Center for Atmospheric Research, 1983).
9. National Center for Atmospheric Research (NCAR), *Annual Report: Fiscal Year 1983* (Boulder, Colorado: NCAR, 1984).
10. National Defence University (NDU), *The Global Impacts of Climate Change in the Year 2000* (Washington, DC: Fort Lesley, J. McNair, 1983).
11. Smit, B., "Climate warming and Canada's comparative position in agriculture", *Climate Change Digest*, CCD89-01, Environment Canada, 1989.
12. Roningen, V.O., "A Static World Policy Simulation (SWOPSIM) modelling framework", *ERS Staff Report No. AGES860625* (Washington, DC: International Economics Division, Economic Research Service, US Department of Agriculture, July, 1986).
13. Reilly, J. (personal communication, 1989).

Chapter 8

1. Pitovranov, S.E., Iakimets, V., Kiselev, V.I. and Sirotenko, O.D., "The effects of climatic variations on agriculture in the subarctic zone of the USSR", in Parry, M.L., Carter, T.R., and Konijn, N.T. (eds), *The Impact of Climatic Variations on Agriculture, Volume 1, Assessments in Cool Temperate and Cold Regions* (Dordrecht, The Netherlands: Kluwer, 1988), pp. 617–722.
2. Squire, G.R., and Unsworth, M.H., "Effects of CO_2 and climatic change on agriculture", *Contract Report to the Department of the Environment* (Sutton Bonnington, UK: Department of Physiology and Environmental Science, University of Nottingham, 1988).
3. Smit, B., "Climate warming and Canada's comparative position in agriculture", *Climate Change Digest*, CCD89-01 (Downsview, Ontario: Environment Canada, 1989).
4. Smit, B., "Implications of climatic change for agriculture in Ontario", *Climate Change Digest*, CCD 87-02 (Downsview, Ontario: Environment Canada, 1987).
5. Jäger, J., *Developing Policies for Responding to Climatic Change*,

Report of two international workshops held in 1987 (Villach, Austria and Bellagio, Italy) (Washington, DC: Environmental Defense Fund, 1988).

6. Salinger, M.J., Williams, W.M., Williams, J.M., and Martin, R.J., (eds), *Carbon Dioxide and Climate Change: Impacts on Agriculture* (New Zealand: New Zealand Meteorological Service; DSIR Grasslands Division; MAFTech, 1990).
7. Yoshino, M., Horie, T,. Seino, H., Tsujii, H., Uchijima, T. and Uchijima, Z., "The effects of climatic variations on agriculture in Japan", in Parry, M.L., Carter, T.R., and Konijn, N.T. (eds), *The Impact of Climatic Variations on Agriculture, Volume 1, Assessments in Cool Temperate and Cold Regions* (Dordrecht, The Netherlands: Kluwer, 1988), pp. 725–868.
8. Balteanu, D., Ozenda, P., Huhn, M., Kerschner, H., Tranquillini, W., and Bortenschlager, S. "Impact analysis of climatic change in the central European mountain ranges", Vol. G., *European Workshop on Interrelated Bioclimatic and Land Use Changes* (Noordwijkerhout, The Netherlands, October, 1987).
9. Santer, B., "The use of general circulation models in climate impact analysis – a preliminary study of the impacts of a CO_2-induced climatic change on Western European agriculture", *Climatic Change*, Vol. 7: 71–93 (1985).
10. Pittock, A.B., "Potential impacts of climatic change on agriculture, forestry and land use" (personal communication, 1989), 5 pp.
11. EPA, *The Potential Effect of Global Climate Change on the United States*. Draft Report to Congress (Washington, DC: US Environmental Protection Agency, 1989).
12. Rowntree, P.R., Callander, B.A., and Cochrane, J., "Modelling climate change and some potential effects on agriculture in the UK", *Journal of the Royal Agricultural Society of England*, November, 1989.
13. Koster, E.A., Dahl, E., Lundberg, H., Heinonen, R., Koutaniemi, L., Torssell, B.W.R., Heino, R., Lundmark, J.E., Stronquist, L., Ingelog, T., Jonasson, L., and Eriksson, E., "Impact analysis of climatic change in the Fennoscandian part of the boreal and sub–arctic zone," Vol. D., *European Workshop on International Bioclimate and Land Use Changes* (Noordwijkerhout, The Netherlands, October, 1987).
14 Williams, G.D.V., Fautley, R.A., Jones, K.H., Stewart, R.B. and Wheaton, E.E., "Estimating effects of climatic change on agriculture in Saskatchewan, Canada", in Parry, M.L., Carter, T,R., and Konijn, N.T.(eds), *The Impact of Climatic Variations on Agriculture, Volume 1, Assessments in Cool Temperate and Cold Regions* (Dordrecht, The Netherlands: Kluwer, 1988), pp. 221–379.
15. Parry, M.L., and Carter, T.R., "The assessment of the effects of climatic variations on agriculture: aims, methods and summary of results", in Parry, M.L., Carter, T.R., and Konijn, N.T. (eds), *The Impact of Climatic Variations on Agriculture, Volume 1, Assessments in Cool Temperate and Cold Regions* (Dordrecht, The Netherlands:

Kluwer, 1988), pp. 11–96.
16. Parry, M.L., Carter, T.R. and Porter, J.H., "The Greenhouse Effect and the future of UK agriculture", *Journal of the Royal Agricultural Society of England*, November, 1989: 120–31.
17. Imeson, A., Dumont, H., Sekliziotis, S., "Impact analysis of climatic change in the Mediterranean region", Vol. F., *European Workshop on Interrelated Bioclimatic and Land Use Changes* (Noordwijkerhout, The Netherlands, October, 1987).
18. Smit, B., Brklacich, M., Stewart, R.B., McBride, R., Brown, M., and Bond, D., "Sensitivity of crop yields and land resource potential to climate change in Ontario", *Climatic Change*, Vol. 14: 153–74 (1989).
19 Bergthorsson, P., Björnsson, H., Dyrmundsson, O., Gudmundsson, B., Helgadottir, A., and Jonmundsson, J.V., "The effects of climatic variations on agriculture in Iceland", in Parry, M.L., Carter, T.R., and Konijn, N.T. (eds), *The Impact of Climatic Variations on Agriculture, Volume 1, Assessments in Cool Temperate and Cold Regions* (Dordrecht, The Netherlands: Kluwer, 1988), pp. 383–509.

Chapter 9

1. Intergovernmental Panel on Climate Change, *The Potential Impacts of Climate Change: Impacts on Agriculture and Forestry* (Geneva and Nairobi: World Meteorological Organisation and United Nations Environment Programme, 1990).

Index

acid deposition, 29
adjoint method, *see* climate impact assessment, adjoint method
adjustments (at farm level) to climatic change, 119-126, 130
  adjustment experiments, 32
  in technology, 30, 104, 117, 121
  in farmed area, 119-120
  in crop husbandry, 126
  in farm management, 104, 117, 124-126
  in land use, 90, 119-124
  see also agricultural policy, crop varieties, irrigation, soil drainage, farm infrastructure
afforestation, 99
Africa, 3, 84, 86, 110-111, 116, 128
  Horn of, 3, 128
  north 19-20
  north-east 19-20
  southern, 19-20, 128
  west, 19, 128
African swine fever *see* diseases
agricultural policy, 30
  see also adjustments, food supply, exports, imports
agricultural potential, 61-87, 92
  changes in, 76-87, 129
  spatial shift of, 122-124, 130
  limits of, 119, 130
  see also agroclimatic indices
agroclimatic indices, 33
agroclimatic zones, 48, 76
Andes, 3, 74
apples, 96
Arabian Peninsula, western, 19-20
Argentina, 6, 86, 104, 108-111, 114-115, 119
  Pampas, 19, 80
Asia, 111
  south-east, 3, 17-18, 20, 60, 108-111, 116, 128
  south-west, 3
Australia, 6, 17-18, 55, 76, 80, 84, 87-88, 96-97, 104, 108-111, 114-116
  eastern, 19
  western, 19, 86, 128

bacterial pathogens, *see* diseases
Bangladesh, 57, 60
barley, 7, 74, 77, 103, 123
  see also yields, barley
barley yellow dwarf virus, *see* diseases
Battelle Institute, 28
beans, 40, 74
biomass production, 27, 75-76
bird cherry aphid (Rhopalosiphum padi), *see* pests
boreal forest, 56
  see also afforestation
Brazil, 7, 108-111, 114-115
  eastern, 19, 20, 128

C_3 plants, 37-41
C_4 plants, 39, 41, 83
cabbage, 40
Canada, 6-7, 16, 91-94, 108-111, 114-116, 123
  Manitoba, 80

Ontario, 80, 94
Prairies, 43, 49, 62, 80, 87, 104, 122-123
Saskatchewan, 44, 72, 88, 91-104
Canadian Climate Centre general circulation model, 14-15, 18
carbon dioxide (CO_2), 9
direct effects of, 37-41, 79, 80, 128-129
in atmosphere, 10, 46
see also Climatic scenarios, GFDL, GISS, OSU, UKMO
carbon monoxide, 11
carcass weight, of lambs, 98-99
of sheep, 98-99
carrying capacity, 86
of sheep on improved grassland, 86, 99
of sheep on unimproved rangelands, 98-99
see also Livestock
cattle, 47, 54, 86, 96
Central America, 3, 20, 108-111, 128
Chile, 86, 104
China, 6, 57, 85, 108-111, 114-116
north and central, 19
climatic scenarios, analogue scenarios, 36, 84
based on GCM outputs, 12, 14-16, 35, 61, 63-68, 91-104
development of, 35-36
IPCC "Business-As-Usual" scenario, 1, 11-12, 16, 18, 57, 60, 70, 74, 76, 130
paleoanalogue scenarios, 84
synthetic, 36, 63, 70
transient, 36, 68
see also General circulation models, Canadian Climate Centre, GFDL, GISS, NCAR, OSU, UKMO
chlorofluorocarbons (CFCs), 9-11
Montreal Protocol, 10
citrus, 123
Climate Impact Assessment Program (CIAP), 24
climate impact assessment, adjoint method, 30-31
direct method, 30-31
impact approach, 24-25
integrated approach, 27-30
interaction approach, 25-27
climatic extremes, 49-51, 70, 93, 128-129
heat waves, 21
see also frequency, of extreme events
coffee, 7
cotton, 40
crop-climate models, 33-34
empirical-statistical models, 33-34
simulation models, 34-35
see also agroclimatic indices, Effective Temperature Sum
crop failure, 50
frequency of, 73
probability of, 50-51, 74
crop varieties, 79
changes in, 120-124
changes to varieties with high thermal requirements, 120-121
changes to more drought-tolerant crops, 121-122
selection of, 131
crop-water availability, *see* soil moisture
cucumber, 40

deforestation, 11
degree days, *see* Effective Temperature Sum
Denmark, 6, 65, 67
diseases, 29, 54-56, 129, 131
control of, 125
African Swine Fever, bacterial pathogens, 56
Barley yellow dwarf virus, 55

fungal pathogens, 55
potato blight, 55, 99
Rift Valley Fever, 54
drought, 25-26, 50, 75, 129
frequency of, 93
risk of, 129

economic models, farm-level, 89
land allocation, 103
national agricultural models, 89
ecosystems, 55-57
Ecuador, 74
Eemian Interglacial, *see* climatic scenarios, paleoanalogue scenarios
Effective Temperature Sum (ETS), 33, 54, 56-57, 65, 79
eggplant, 40
Egypt, 6, 57, 60
El Niño/Southern Oscillation (ENSO), 97
employment, 91-92
Europe, 111
Alps, 74
northern, 80, 83, 110, 114-116
Low Countries, 104
Pyrenees, 74
southern, 17-18, 87, 110, 124, 128-130
western, 19, 124
European Community, 7, 108-111, 114-116
European Commission, 27, 35
European corn borer (ostrinia nubilalis), *see* pests
evapotranspiration, 12, 16, 48, 63, 75, 85, 93, 113, 128

farm income, 92, 100
farm infrastructure, 8
changes in, 125-126
farm management, *see* adjustments
fertilizer application, 79, 131
changes in, 125-126
Fescue, 40
Finland, 72, 77-79, 88, 99-101
first-order interactions, 26-28
see also disease, droughts, livestock, pests, second-order interactions, soils, third-order interactions, weeds, yields
flooding, 2, 37, 60
Food and Agriculture Organization (FAO), 3
food deficits, 39
exports, 6, 106-111, 128
imports, 6, 106-111
prices, 6, 106-111
security, 6, 7, 105-118, 127
stocks, 95-96
supply, 105-118, 131
France, 6-7, 16
frost, 43, 49
frequency, of crop failure, 73
of extreme events, 20, 85
fruit, 65
production, 94
stone fruits, 96
fungal pathogens, *see* diseases

general circulation models (GCMs), 11-12, 16
see also Canadian Climate Centre, GFDL, GISS, NCAR, OSU, UKMO
Germany, 65
GFDL (Geophysical Fluid Dynamics Laboratory) general circulation model, 13-15, 18-19
GFDL 2 × CO_2 scenario, 67, 80-82, 93-94
GISS (Goddard Institute for Space Studies) general circulation model, 13, 19
GISS 2 × CO_2 scenario, 52, 63-64, 67-68, 72, 79-83, 86, 93-94, 99, 101
GISS transient response scenario A, 69-70

grain aphid (sitobian avenae), *see* pests
grain maize, *see* maize
greenhouse effect, 9-23
see also greenhouse gases, climatic scenarios
greenhouse gases (GHG), 1, 9-10
see also Carbon dioxide, Chlorofluorocarbons, Methane, Nitrous Oxide
groundwater *see* salinity
growing season, 19, 43-44, 63, 65, 70, 83

hay, 123
see also yields, hay
heat stress, 129
risk of, 51
heatwaves, *see* climatic extremes
holocene optimum, *see* climatic scenarios, paleoanalogue scenarios
horn fly, *see* pests
horticultural crops, 97

Iceland, 55, 72, 86, 88, 97-99, 125
IIASA/UNEP project, 29, 31, 72, 78-80, 85, 91-104, 109
impact assessment, *see* climate impact assessment
impacts, *see* first-order interactions, second-order interactions, third-order interactions
India, 2, 39, 108-111
Indonesia, 57, 108-111
interactions, *see* first-order interactions, second order interactions, third-order interactions
International Federation of Institutes of Advanced Study (IFIAS), 26
Intergovernmental Panel on Climate Change (IPCC), 1, 17, 105, 127
see also Climatic scenarios, IPCC "Business-As-Usual"
Intertropical Convergence Zone (ITCZ), 75
Iran, 6
Iraq, 6
irrigation, 94, 131
changes in, 125-126
Italy, 6

Japan, 6, 68, 88, 94-96, 114-115
Hokkaido, 65, 72, 77, 80, 95, 121
northern, 79, 119
Tohoku, 72, 77, 94-95
jirrison weed, *see* weeds

Kenya, 48, 86
Korea, Republic of, 6

lambs, 47, 86, 98,
see also carcass weight, livestock
land degradation, *see* soil erosion
lettuce, 40
livestock, 46-47, 86
feed requirements for, 98
see also carcass weight, carrying capacity, cattle, lambs, pigs, poultry, sheep
locust, *see* pests

Maghreb, 3, 128
maize, 7, 38-39, 62-63, 65, 89, 93, 97, 102-103, 112-118, 122, 127, 129
grain maize, 54, 69-71
silage maize, 39, 71, 123
see also yields
Maldive Islands, 60
Marginality, 7
of agriculture, 7, 74, 76, 130
of farmers, 7
economic marginality, 7
social marginality, 8
Mars, 9

methane (CH_4), 9-10
Mexico, 6
millet, 39, 129
monsoon, *see* precipitation
moorland, 74

National Defense University (NDU), 25
NCAR (National Center for Atmospheric Research) general circulation model, 13, 19
net primary productivity, *see* biomass production
New Zealand, 6, 55, 65, 87-88, 96-97, 104, 119-120
nitrous oxide (N_2O), 9-10
nomadic herding, 26
North America, 19, 130
  central, 17-18

oats, 51, 73, 103, 123
  see also yields
Oceania, 110-111
okra, 40
oilseeds, 101
olives, 123
OSU (Oregon State University) general circulation mode, 13
  OSU 2 × CO_2 scenario, 67

Pakistan, 114
Palmer Drought Index, 33
pears, 96
pests, 29, 54-56, 129, 131
  control of, 125
  bird cherry aphid (Rhopalosiphum padi), 55
  European corn borer (Ostrinia nubilalis), 53-54
  grain aphid (sitobian avenae), 55
  horn fly, 54
  locust, 55
  potato leafhopper, 54
  ticks, 55
photosynthesis, 37-41
pigweed, *see* weeds
pigs, 47
plant breeding, *see* crop varieties
Pliocene optimum, *see* climatic scenarios, paleoanalogue scenarios
potatoes, 74, 102-103, 123
  see also yields
potato blight, *see* diseases
potato leaf hopper, *see* pests
poultry, 47
precipitation, decrease in, 17-18, 26, 45, 83
  increase in, 2, 12, 16-18, 45, 83-84
  reliability of, 48
  variability of, 48, 97
probability, of crop failure, 50-51, 74
  of extreme events, 50
  see also frequency
production costs, minimizing, 103
profitability, 101

ragweed, *see* weeds
rainfall, *see* precipitation
rangelands, 39, 86
  unimproved, 99
  see also carrying capacity, livestock, yield
rice, 38, 40, 65, 68, 89, 94, 112-118, 127
  stocks of, 94-96
  yield index of, 95
  see also yields
Rift Valley fever, *see* diseases
risk, levels of agricultural, 49-50, 72
runoff, 27
rye, 101
  winter, 102

Saudi Arabia, 6
Sahel, 2-3, 7, 17-18, 25, 75
salinity, of groundwater, 37, 60

of soils, 125, 131
Scandinavia, 48, 78, 104, 122
sea level rises, 37, 57, 60
second-order interactions, 26-28
see also employment, first-order interactions, food prices, food security, food stocks, second-order interactions
Senegal, 75-76
Sheep, 47, 86, 96-98
wool, 86, 97
see also carrying capacity, lambs, livestock
sicklepod, *see* weeds
silage maize, *see* maize
soils, 103
drainage of, 125
erosion of, 2, 25, 29, 52-54, 91, 125, 131
fertility of, 52, 125
microbes in, 131
porosity of, 48
see also salinity
soil moisture, 2, 14-16, 18-20, 25, 47-49, 63, 75, 83, 128, 130
sorghum, 7, 39, 89, 129
South Africa, 6, 108-111
South America, 3, 110, 116, 128
soya beans, 7, 38, 93, 112-118, 122, 127
see also yields
spring wheat, *see* wheat
Static World Policy Simulation (SWOPSIM), 113-118
storms, 20
sugar beet, 40, 101
sugarcane, 39
sunflower, 40

technology, *see* adjustments
temperature increase, 12-13, 16-18, 21-22, 41-48
threshold temperatures, 57
see also climatic scenarios, Effective Temperature Sum
Thailand, 6, 57, 108-111
third-order interactions, 26-28
see also employment, first-order interactions, food prices, food security, food stocks, food supply, second-order interactions
ticks, *see* pests
tidal penetration, *see* salinity
timing of farm operations, changes in, 126
Toronto Conference on the Changing Atmosphere, 105
Transpiration, 41
rates of, 12

UKMO (United Kingdom Meteorological Office) general circulation model, 13-15, 18
UKMO 2 × CO_2 scenario, 83
United Kingdom, 6, 65, 71-73, 104
United Nations Environment Programme (UNEP), 1, 28
U.S. Environmental Protection Agency (EPA), 88-89
USA, 6-7, 16, 25, 50, 81-82, 89-91, 108-111, 114-116, 129
Alaska, 119
Appalachia, 89
California, 83
Corn Belt, 7, 45, 50, 52, 54, 62-63, 87
Great Lakes region, 83, 89, 120
Great Plains, 7, 16, 49, 83, 87, 89, 123
Louisiana, 57
southern, 19, 128-129
USSR, 6, 25, 52, 83, 88, 101-104, 108-111, 114-116, 125
Cherdyn region, 72
Kazakhstan, 130
Leningrad region, 72, 80, 101
Moscow region, 77, 101, 122
northern, 79, 123
Russia, central, 65

Siberia, 19
south European, 87
Soviet Central Asia, 19
Ukraine, 45, 49, 80, 130
Ural mountains, 80, 101
Volga Basin, 49

vegetables, 65, 102, 123
production, 94
vegetation zones, 57-59
vines, 123
vulnerable regions, 3-6, 19-20, 128

water use efficiency, 41
weeds, 39, 129
C_3 weeds, 40
control of, 125-126
Jirrison weed, 40
pigweed, 40
ragweed, 40
sicklepod, 40
wheat, 7, 38, 40, 43, 64, 65, 101, 112-118
spring wheat, 63, 102, 123
winter wheat, 64, 103, 123
see also yields
wind erosion, 92-93
windspeed, 48
winter chilling (vernalization), 51-52
winter wheat, *see* wheat
World Meteorological Organization (WMO), 1

yields, 49, 61, 76-87
barley, 44, 78, 80, 99-100
grain maize, 80
grassland, 86
green vegetables, 80
hay, 44, 80, 97-99
maize, 49, 50, 80-81, 83, 85
milk, 86
oats, 78, 80, 100
pototoes, 80
quality, 41
rice, 121
soybeans, 80, 83
spring wheat, 44-45, 49, 80, 91-94, 99-101
wheat, 49-50, 82-83
winter wheat, 45-46, 80